To: Professor David G. Shappiro
with Complements
from Bron Cymborowski

Warszawa 28.04.1993

Insect Endocrinology

Insect Endocrinology

Bronisław Cymborowski

Institute of Zoology, University of Warsaw, Poland

ELSEVIER
Amsterdam – Oxford – New York – Tokyo
PWN – POLISH SCIENTIFIC PUBLISHERS
Warszawa
1992

Revised and updated translation of the Polish edition *Endokrynologia owadów*, published by Państwowe Wydawnictwo Naukowe, Warszawa 1984

Translated by Maria Radziwiłł

Cover design by *Zygmunt Ziemka*

Distribution of this book is being handled by the following publishers:

For the USA and Canada
ELSEVIER SCIENCE PUBLISHING CO., INC.
655 Avenue of the Americas, New York, NY 10010

For Albania, Bulgaria, Cuba, Czecho-Slovakia, Hungary, Korean People's Democratic Republic, Mongolia, People's Republic of China, Poland, Romania, the USSR, Vietnam and Yugoslavia
PWN — Polish Scientific
Publishers, Miodowa 10,
00-251 Warszawa, Poland

For all remaining areas
ELSEVIER SCIENCE PUBLISHERS B.V.
Sara Burgerhartstraat 25
P.O. Box 211, 1000 AE Amsterdam, The Netherlands

Library of Congress Cataloging in Publication Data

Cymborowski, Bronisław
[Endokrynologia owadów. English]
Insect endocrinology / Bronisław Cymborowski : [translated by Maria Radziwiłł].
p. cm.
"Revised and updated translation of the Polish edition
Endokrynologia owadów" — — T.p. verso.
Includes bibliographical references and index.
ISBN 0-444-98704-5
1. Insects — Physiology. 2. Insect hormones. 3. Insects —
Reproduction — Endocrine aspects. 4. Endocrinology. I. Title.
QL 495. C 9613 1991
595.7'0142—dc 20
ISBN 0-444-98704-5

Printed in Poland by D.N.T.

Contents

In commemoration of Professor
Stefan Kopeć
pioneer in insect endocrinology

Preface

The problems of development and metamorphosis of insects have long been objects of interest to biologists. They have searched above all for the factors directly controlling such dramatic transformations in the insect's life as the transition from the larval stage to the pupa, and from the pupa to the imago which leads quite a different kind of life from the former.

In this line of research, the investigations of the Polish scientist Stefan Kopeć played a prominent role. The subject of his doctoral thesis (1912) was insect metamorphosis, and he thus laid the foundations of the new discipline in biological science that is insect endocrinology. In his thesis, published by the Polish Academy of Science in Cracow in 1917 under the title "Experimental studies on the metamorphosis of insects", he demonstrated the role of the insect brain as an endocrine gland. In his summary he states: "The brain plays an exceptional role in relation to the general processes of insect metamorphosis, and its presence is indispensable for initiating the transformation processes. The nature of its influence in this respect is most probably chemical; therefore, the brain should be considered as an endocrine gland (production of tyrosinase ?)".

In the same work Kopeć introduced the notion of a critical period in the life of the larva, associated with release of an appropriate amount of the brain hormone into its haemolymph. He expressed it as follows in his publication: "For a certain definite time before pupation of the caterpillar the amount of substances released by the brain is already sufficient for the caterpillar to undergo complete transformation, even if deprived of the brain at this moment." It is noteworthy that this conclusion was advanced on the basis of very simple experiments in which Kopeć, for the first time, applied ligatures at various levels of the larval body, and his method is widely used to this day in insect endocrinology.

The results of the experiments carried out by Kopeć were soon confirmed in morphological and histological investigations. Many endocrine centres and glands were detected. Endocrinology began gradually to accumulate diverse data concerning the structure and distribution of the particular organs of the endocrine system of these animals, and this type of research has since been continued. There has been, however, a marked transition from classical morphological studies to studies of a physiological and biochemical nature. This has led, among other things, to the possibility of the use of insect hormones and their analogues in studies of insect control. This is particularly important at the present time, when many initially efficacious plant protection agents (e.g. DDT) have failed and the necessity has arisen of searching for new, above all biological, ways of controlling mass outbreaks. Insect hormone analogues seem to meet many important requirements of contemporary insect control. The increasing possibilities for their practical application are no doubt the cause of the wide interest of biologists in insect endocrinology. Today it has become a very wide sphere of activity in the biological sciences, and coverage of all its aspects is outside the scope of this book.

After briefly acquainting the reader with the history of research on the insect endocrine system (Chapter 1), the centres and glands of this system are described. Particular attention is devoted to the neurosecretory cells of the brain and some of the nerve ganglia. Much space is devoted to neurohaemal structures, with special stress on their role in the functioning of the insect endocrine system. Since the methods used in investigations on insect endocrinology differ widely from those commonly used for other animal groups, it was decided to present only the most important methods, the use of which has greatly extended and continues to extend our knowledge about this group of animals (Chapter 3).

The most extensive part of this book is the chapter on the hormonal regulation of insect growth and metamorphosis (Chapter 4), which contains recent results on the function of the endocrine system. Particular attention is devoted to the mechanism of action of various hormones. The chapter considers problems of cell and tissue differentiation in the developing insect in the presence or absence of specific hormones in the haemolymph, particularly juvenile hormone and ecdysone. This chapter ends with a discussion on the mechanisms of interaction of hormones during growth and metamorphosis, with special reference to the functioning of the organism as a whole.

In the succeeding chapters, the participation of insect hormones in processes such as reproduction, diapause and the regulation of metabolism and behaviour are discussed. Finally, prospects for the insect hormones as insecticides are considered.

Insect pheromones, although very intensively studied, are not considered in this book. They are produced for "export" and play an important role in chemical communication between particular individuals of insect populations. The possibility is stressed, however, that pheromones influence the functioning of the insect endocrine system.

The extensive reference lists following the text of each chapter should encourage the reader to undertake independent studies in the extremely interesting field of insect endocrinology.

B. Cymborowski

Chapter 1

Introduction to insect endocrinology

1.1. Outline history

Research on the hormonal regulation of various vital processes in invertebrates started relatively late, and scientists throughout the world were impressed by the first experimental results in the field of human endocrinology.

The experiments of Brown-Sequard who in 1889 used extracts of sex glands to rejuvenate the body became famous. The now classical experiments followed, consisting of the removal and transplantation of endocrine glands from various animals and the investigation of their effects. It is not known precisely when and under what conditions the concept of the chemical integration of the body arose. The term "hormone" (from a Greek word meaning to stimulate or set in motion) was used for the first time by the English physiologist Ernest H. Starling in 1905. This is not entirely accurate, as it is now known that hormones can not only stimulate but can also inhibit certain processes, thus establishing a correct balance in the various metabolic transformations occurring in the body.

Similarly, the introduction some years later of the expression "endocrinology" (from Greek *endon*, inside and *krinein*, to secrete) does not correspond exactly with the present scope of studies in this sphere of biological science. Endocrinology in general may be defined as the science dealing with the chemical integration of the body, in which regulation of vital processes occurs under the influences of substances produced in definite endocrine glands. If, however, we start to analyse plants and animals in detail, it appears that diverse chemical types of integration exist which lie well outside the scope of endocrinology. A classical example here is the action of the so-called "queen substance" produced by the queen bee, which enters the workers' bodies via their food and has an inhibitory effect on the development of their ovaries (Butler, 1954). This would therefore be an ectohormone—a substance produced for "export", as in the case of pheromones, which serve to attract the opposite sex. Should investigations of this type be considered as belonging to the field of endocrinology or not? Moreover, practically all cells have secretory properties and release various products of metabolism into the body. This makes a precise definition of an endocrine gland impossible.

There is another very important problem in that until now research on the regulation of the vital processes of the body has been conducted along two independent lines: chemical integration studied by endocrinologists, and nervous integration investigated by neurophysiologists. It is known at present that both these systems interact closely, and there are thus a number of problems at the border between endocrinology and neurophysiology; frequently it is not clear in which discipline they should be classified. Some groups of nerve cells secrete what are called neurohormones, which may be transported via axons to remote target tissues where they exert their specific effects. Such cells are also able to conduct electric potentials, that is, they behave like normal neurons. In addition, the axons of such cells may have their endings in the neurohaemal organs, which temporarily accumulate the chemical substances produced and then release them directly into the circulatory system. Thus, the question arises in which of the above-mentioned cases are we dealing with an endocrine organ? Or are they both endocrine organs?

These problems and many others became particularly serious when the scope of endocrinological research expanded, and scientists, no longer limited themselves to gaining a knowledge of the mechanisms underlying pathological changes and their treatment. In the 1920s endocrinologists limited their interests to "freaks of nature" such as abnormal growth, hermaphroditism etc.

Luckily, however, endocrinology was guided along the right path, with the help of the Polish researcher Stefan Kopeć who, using very simple methods, demonstrated that a complex system of chemical integration of the body exists not only in vertebrates but also in invertebrates.

The now classical investigations of Kopeć (1917, 1922) were performed on larvae of the gypsy moth (*Limantria dispar*), common in Poland. He proved that metamorphosis of this insect is hormonally regulated. This conclusion was based on the results of experiments in which insect larvae were ligated at various developmental stages. When larvae of the last instar were ligated between the head and the remaining part of the body, the adult moths that appeared about 10 or more days after the last larval moult were normal in appearance except for the absence of the head. On the other hand, when the ligature was applied

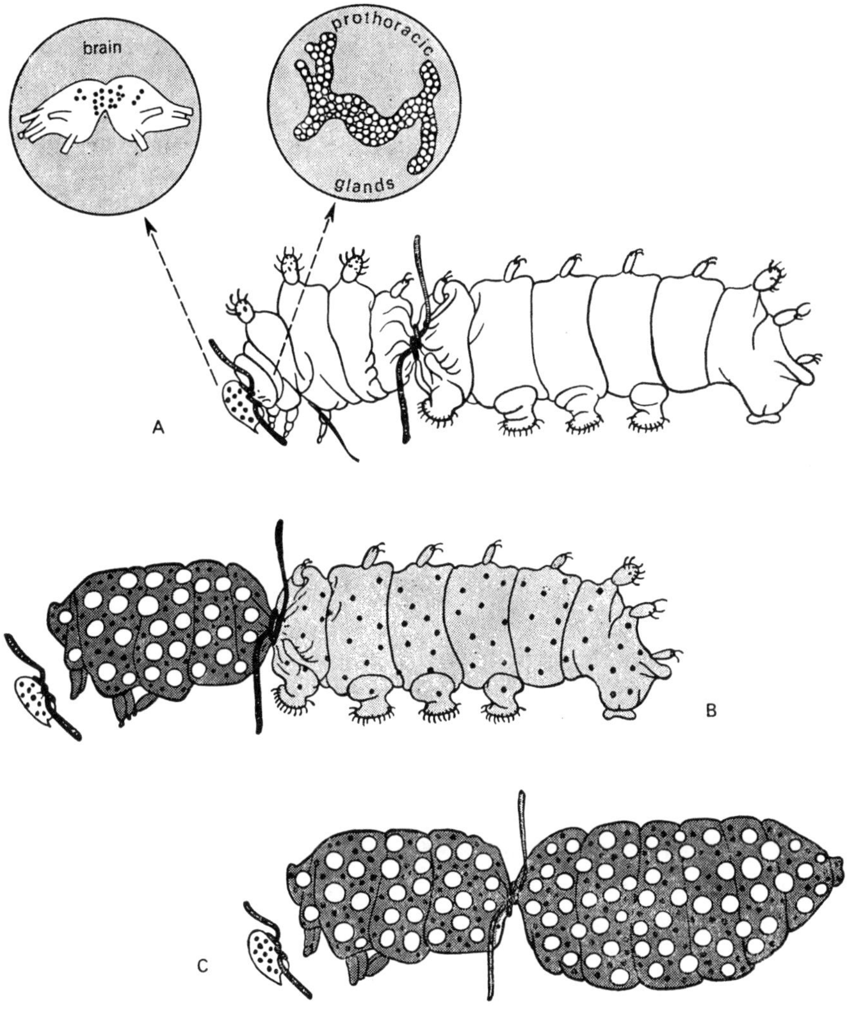

Figure 1.1. Experiments on hormonal regulation of insect metamorphosis using ligatures. After Williams (1950).

A—ligation of larva behind head capsule deprives the rest of the body of the influence of the brain hormone (PTTH—dots), while ligation behind the thorax also isolates the prothoracic glands. B—ligation after PTTH release by the brain, but before ecdysone release. Thus, pupation occurs only in the anterior part of the body, in which the prothoracic glands are stimulated by PTTH to produce ecdysone (circles). C—ligation after PTTH has stimulated the prothoracic glands to secrete ecdysone. In spite of ligation both parts of the larva pupate.

before the last larval instar, pupation did not occur, and the insects remained in a larval state for a prolonged period. The same author believed that in the larval life of *Limantria dispar* there is a critical period after which transformation of the larva to the pupa and then to the imago occurs. In further experiments Kopeć noted that if, before the critical period, a ligature is fitted at the mid-point of the body of a final-instar larva, so that the haemolymph of the anterior part of the body does not mix with that of the posterior part, then only the anterior part, containing the brain, undergoes pupation (Fig. 1.1). When, however, the ligature is applied after the critical period, then both the anterior and the posterior part undergo metamorphosis.

On the basis of these results Kopeć suggested that the brain releases a substance into the haemolymph which plays a decisive role in the transformation of the larva into a pupa.

The experiments described by Kopeć brought about a flood of investigations into the hormonal regulation of vital processes in insects, so that today we have a number of monographs reviewing the results of experimental studies (Wigglesworth, 1954, 1970; Gersch, 1964; Jacobson, 1965; Novak, 1966; Butler, 1967; Highnam and Hill, 1969; Beroza, 1970; Jenkin, 1970; Bowers, 1971; Slama *et al.* 1974; Kerkut and Gilbert, 1985).

1.2. Fundamental problems of insect endocrinology

What prompted researchers the world over to investigate insect hormones on so large a scale ? As in other fields of science, curiosity, the desire to reveal the mechanisms controlling the complex processes of growth and metamorphosis of insects formed the basis of this intensive research effort. With the founding of the discipline of comparative endocrinology, there arose a new possibility for the investigation of the mechanism of hormone action. There was an urgent need to understand the evolutionary development of these mechanisms, and insects proved to be very convenient experimental subjects for this purpose. This was mainly because (1) they can be reared on a large scale in the laboratory, (2) various drastic experiments can be performed on them, and (3) they are in many respects simpler systems than vertebrates, thus allowing easier understanding of the chemical regulation of many of their vital processes. Among the latter, metamorphosis occupies a unique place, comprising the whole cycle of transformations leading through successive larval stages to the adult form in the *Hemimetabola* (incomplete development) and through larval stages and pupa to the adult form in the *Holometabola* (complete development). Insects thus proved to be convenient material for the investigation of biological processes in general, such as growth, cell differentiation, heredity, metabolism and many others. Moreover, the endocrine system of insects is so situated in the organism that its components can be removed without destroying other tissues and systems. This makes possible the study of physiological processes such as metamorphosis, diapause and reproduction.

Insect hormones are the object of biochemical investigations not only from the point of view of their chemical structure, but also and perhaps above all, as regards their mechanism of action, as for instance in the process of gene activation.

Another fundamental aspect of insect endocrinology to which Novak (1971) draws attention is the possibility of utilizing the results in applied entomology. Research indicates that this is possible for two reasons:

1. a knowledge of the mechanisms of the endocrine system makes it possible to predict the mass appearance of pests and to develop methods of control using insecticides,

2. recognition of the chemical structure of insect hormones and the development of methods for their synthesis, as well as the elucidation of their mechanisms of action, permits their utilization in insect control, as had been suggested by Williams (1956).

It is believed that insects will be unable to develop resistance to these substances, as the compounds concerned are indispensable for normal growth and metamorphosis. These problems will be treated extensively in Chapter 9 of this book.

1.3. Contemporary problems of insect endocrinology

There is at present in insect endocrinology a tendency for morphological investigations to give way to studies of a physiological and biochemical nature. This does not mean of course that all morphological problems have now been exhaustively studied and have been resolved. In the case of insects, however, the classical assertion that there is a distinct correlation between the structure of an organ and its function is not always true; this will be discussed in further chapters of this book. The necessity has arisen, therefore, of creating new models of hormone activity, based above all on chemical and biophysical foundations.

We now know that all nerve cells transmit their information by releasing special chemical substances which act upon other cells. The transmitted informa-

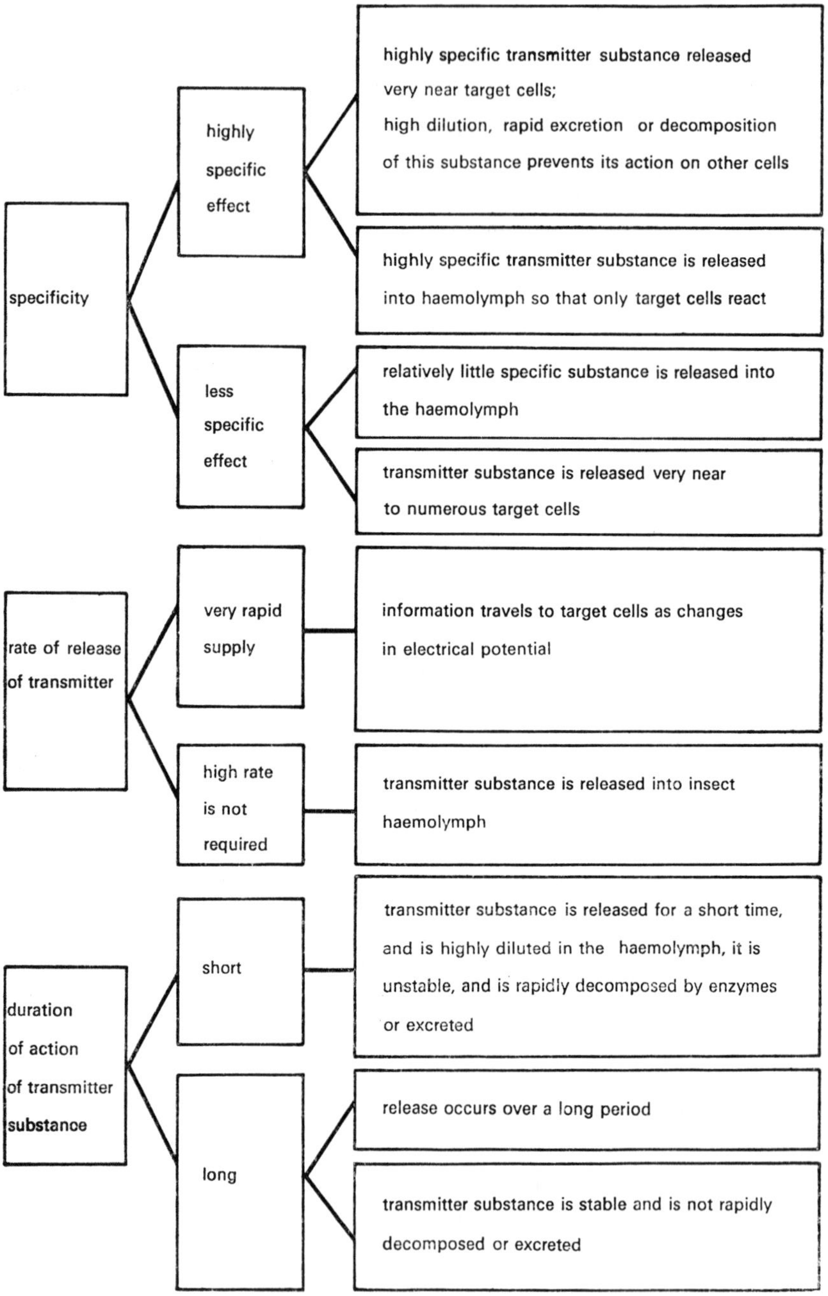

Figure 1.2. Possible means of transmission of information to the target cells and tissues by the insect nervous system.

tion may be specific (reaching only one cell) or general (reaching many cells). In the insect body, as in vertebrates, such information may be transmitted either very rapidly or slowly; it may be required to remain for a short time in the body, or to subsist for an extended period at a particular level. The nervous system of insects is so differentiated that it fulfils all the above requirements: high specificity, different velocities of information transmission, and maintenance of the transmitter substance in the haemolymph over the required period (Maddrell, 1974). These different possibilities of transmitting information by the nervous system are diagramatically presented in Figure 1.2. As seen in the figure, cells receiving the information are responsible for modifying information in a specific way.

If we consider how many problems are involved in the functioning of the insect nervous system and the endocrine system that are intimately bound up with it, it will be seen that it is very difficult to interpret experimental results concerning only insect endocrinology.

Later in this chapter, the main problems that are the focus of interest of numerous researchers will be presented.

1.3.1. Insect metamorphosis

Hormones regulating insect metamorphosis are also called growth and differentiation hormones. At present, however, there is increasing experimental evidence indicating that there are no hormones in insects that simultaneously stimulate growth of all tissues of the body. For instance, activation hormone (AH), otherwise referred to as the cerebral hormone (prothoracicotropic hormone, PTTH), and considered to be the typical growth hormone, does not stimulate growth of the follicular cells of the oocyte epithelium of mature *Diptera* and *Hemiptera*. Follicle growth in these insects is stimulated by the hormone derived from the *corpora allata* (*CA*), that is juvenile hormone. Many stimulatory effects also occur in the insect in the absence of moulting hormone (ecdysone) released by the prothoracic glands (Chapter 3).

1.3.2. Target tissues for insect hormones

The selective action of insect hormones raises the problem of target cells, that is cells sensitive to only one hormonal factor and not susceptible to others. The target cells usually respond to the action of the hormone in question by changes in their structure and metabolism, implying changes in specific physiological processes.

It is still difficult at present to establish which tissues and cells are the specific target organs of particular insect hormones. It has, however, been found on the basis of an increasing number of reports of results that for the activating hormone, for instance, cells present in other endocrine centres such as the *corpora allata*, prothoracic glands, Malpighian tubules and fat body, as well as other tissues not included in the endocrine system, may be target cells. For the hormone produced by the *CA*, and for ecdysone, tissues such as the epidermis and imaginal discs (primordia of adult organs) are targets. In the absence of ecdysone, the hormone produced by the *CA* has as its target tissue the epithelial cells of the ovarian follicles.

These few examples are sufficient to demonstrate that the problem of target organs is extremely complex. Moreover, if we consider the fact that organs may change their sensitivity to different hormones during development, the whole matter becomes still more complicated.

The ability of cells, to react to the effect of a particular hormone seems to be determined genetically, and is associated with gradual differentiation within the tissues. As observed in vertebrates, individual hormones in insects may stimulate growth and metabolism in one target tissue without exerting any influence on others.

The experimental difficulties involved in attempts to ascertain the target cells for particular insect hormones do not at present allow generalized theories about the mechanism of their action to be put forward.

1.3.3. Direct and indirect action of hormones

As in vertebrates, in insects hormones evoke a whole chain of physiological changes in the body. Target cells responding to a particular hormone may produce and release hormonal substances which act on other tissues. The question therefore arises of direct and indirect action of hormones. In insects it is very difficult to determine mechanisms of direct action because hormone action usually involves numerous morphological and biochemical changes within the target cells. Such an action of the hormone should therefore be considered as indirect.

From a theoretical point of view we may consider a situation such as that occurring when the target cells, under the influence of a given hormone, produce a substance acting on other cells. Then all reactions are the result of indirect action of this hormone, but at the same time there may be the result of direct action of this intermediary substance. In such a case we could probably distinguish first- and second-order target tissues when describing the whole reaction chain.

An example of direct action of a hormone on cells is the influence of allatotropic hormone on the *corpora allata* (Muszyńska-Pytel, 1987). The mechanism of its action is not known, but its effect is on cell growth, activation of respiratory enzymes and enhanced RNA synthesis, and this subsequently leads to the synthesis and release of juvenile hormone. The latter acts in turn on the cells of the follicular epithelium of the oocytes, inducing changes similar to those brought about by allatotropic hormone in the cells of the *CA*. Thus, allatotropic hormone exerts a direct influence on the *CA* and indirectly influences the oocyte follicles of insects.

1.3.4. Feedback mechanisms

Recent research has demonstrated that the complete synchronization of the endocrine insect system may be achieved through special feedback mechanisms, homeostatic reactions and factors limiting particular functions (Wigglesworth, 1964). An example of feedback is the stimulation of protein synthesis in epithelial cells of the insect gut by activating hormone. Stimulation occurs only when undigested food is present. The amount of the latter limits the rate of enzyme synthesis, which lasts as long as both factors are available. Another example is the stimulatory action of juvenile hormone upon the oocyte's follicle cells. This occurs only when there is a certain minimal amount of the material necessary for yolk formation in the fat body or in the haemolymph. If such a reserve is lacking the stimulatory effect does not occur, although activating hormone may be directly influencing the *corpora allata*. This will result in release of juvenile hormone into the haemolymph.

These two examples are sufficient to show that, in order to initiate a chain of hormonal reactions in the body, appropriate stimuli from the external or the internal environment of the insect are indispensable. Neurosecretory cells in the brain are the first link of this chain. They secrete their products, however, only when they are stimulated by the exteroreceptors or interoreceptors (Maddrell, 1974). As regards the homeostatic mechanism responsible for the maintenance of a particular substance at a constant level, these receptors have to measure constantly the concentration of the substance and, when necessary, switch its synthesis on and off (Fig. 1.3). Steel (1973) supplied convicing evidence for negative feedback between the neurosecretory cell activity of the heteropteran *Rhodnius prolixus* and the level of moulting hormone (ecdysone) in the haemolymph. When haemolymph from an older insect containing ecdysone was introduced into a younger insect, the neurosecretory cells in the younger insect's

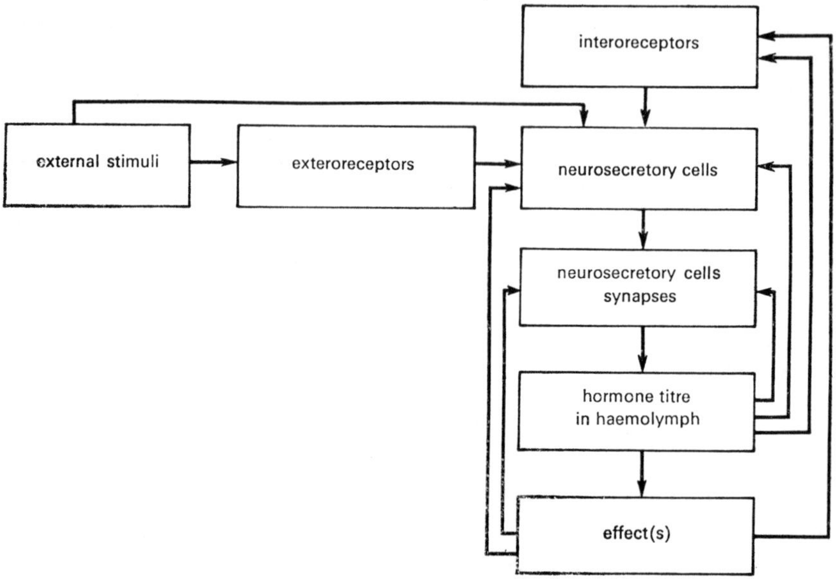

Figure 1.3. Feedback mechanisms between the nervous system and the endocrine system of the insect.

brain were found to be similar to those observed in older individuals in that the production and release of neurosecretion was inhibited. There are probably numerous such feedback mechanisms in insects, but so far data from this field are lacking, or the results have not been convincing.

The existence of homeostatic mechanisms and intermediate reactions in the insect body raises the possibility of exploiting analogues of hormones in insect control. If we realize that a hormone may act under the influence of different loci in the DNA, it may be expected that any substance blocking or releasing the information contained in the DNA may severely disrupt its hormonal effect. From a theoretical point of view, this mimetic action of the substance, similar to that of the hormone, will be the stronger the more intermediate stages exist in the hormonally regulated reaction. An example of this is the action of cholesterol or hyaluronic acid, analogous to that of activating hormone.

1.3.5. Hormonal regulation of insect metabolism

Numerous scientists now suggest that insect hormones produce a change in metabolism, which is mainly manifested in a change in oxygen uptake. The direct effect would consist in stimulation of nucleic acids and protein synthesis, a change in the permeability of cell membranes, etc. No doubt insect hormones regulate the main biochemical processes in insects by their influence on growth and metamorphosis. The range and diversity of the biochemical changes evoked by these hormones indicate that perhaps there are a number of substances of a hormonal nature which have not yet been detected, and which may exert a direct influence on many biochemical processes.

1.3.6. Molecular aspects of insect endocrinology

When considering the effect of insect hormones, it may be wondered how it is possible that one and the same substance causes so many diverse changes in the functioning of the tissues of the body. The above-mentioned activating hormone, produced by the insect brain, may induce juvenile hormone synthesis by the *corpora allata*, ecdysone synthesis by the cells of the prothoracic glands, and synthesis of specific proteins in the fat body cells and many other cells. We cannot understand these facts without taking into account genetic factors and specific differences within particular tissues. A general theory is currently being advanced claiming that hormones act, via the cellular genome, as repressors and derepressors of genetic information (Kroeger, 1971). On the other hand, cells of different tissues react to hormones in a manner that is specific to the particular tissue, and is genetically programmed.

1.3.7. Diurnal changes in the functions of the insect endocrine system

In recent years much literature has been accumulated on the diurnal variation in the functioning of the endocrine system, both in vertebrates and invertebrates. This variation is connected with the existence of the so-called biological clock (Cymborowski, 1976). The functioning of this clock is manifested in the form of diurnal changes in physiological processes and the behaviour of the animal. Circadian rhythms have a place of their own in insect development (Cymborowski, 1974). It appears that practically all stages of insect development—hatching, moulting, pupation and the eclosion of the adult from the pupa—occur at strictly defined times, called gates, within the 24-hour day. This is evidently connected with rhythmic production and release into the haemolymph of a number of hormones participating in the regulation of these processes.

The basic problems in contemporary insect endocrinology are, therefore, those concerned with the mechanisms of the rhythmic changes in the functioning of particular endocrine glands. A knowledge of these would be of practical importance, particularly in the control of pests.

It has recently been revealed that most rhythmic processes occurring in insects are of an endogenous nature, i.e. they undergo characteristic diurnal changes independently of environmental factors, which merely synchronize these processes. Consequently, many researchers stress the genetic aspects of the circadian rhythms of insects (Konopka and Benzer, 1971), and investigations of these rhythms involve a number of problems at the molecular level.

1.4. Summary

From the moment when Kopeć established that the process of metamorphosis in insects is regulated hormonally, research in insect endocrinology began an eventful development. The first steps along this line of development were morphological investigations, leading to a knowledge of the structure of the insect endocrine system. Even the first morphological studies were frequently associated with strictly physiological ones, so that it was possible to ascertain not only the structure of particular glands of the endocrine system, but also their function.

At present, research of a morphological nature is giving way to physiological-biochemical studies. An immense amount of data has been accumulated in the field of insect endocrinology, and this requires the creation of new models of hormone action, based above all on chemistry and biophysics.

Individual chapters of this book will devote special attention to contemporary problems of insect endocrinology, both those related to morphology and, above all, those related to the biochemistry and physiology of insects.

1.5. References

Beroza M. 1970. *Chemicals Controlling Insect Behavior*. Academic Press, New York.

Bowers W. S. 1971, Insect hormones and their derivatives as insecticides. *Bull. World Health. Org.*, **44**; 381–389.

Butler C. G. 1954. The method and importance of the recognition by a colony of honey-bees (*Apis mellifera*) of the presence of its queen. *Trans. Roy. Entomol. Soc.*, **105**; 11–29.

Butler C. G. 1967. Insect pheromones. *Biol. Res.*, **42**; 42–87.

Cymborowski B. 1974. Rytmy dobowe w rozwoju owadów [Circadian rhythms in insect development]. *Kosmos* **B 3**; 231–245.

Cymborowski B. 1976. *Zegary biologiczne* [Biological clocks]. PWN, Warszawa.

Gersch M. 1964. *Vergleichende Endokrinologie der wirbellosen Tiere*. Akademische Verlags Gesellschaft, Leipzig.

Highnam K. C. and Hill L. 1969. *The Comparative Endocrinology of the Invertebrates*. E. Arnold Ltd., London.

Jacobson M. 1965. *Insect Sex Attractants*. Wiley, New York.

Jenkin P. M. 1970. *Control of Growth and Metamorphosis*. Pergamon Press, Oxford.

Kerkut G. A., Gilbert L. I. (eds.). 1985. *Comprehensive Insect Physiology, Biochemistry and Pharmacology*, vol. 1–12. Pergamon Press, Oxford, New York, Toronto, Sydney, Paris, Frankfurt.

Konopka R. J., Benzer S. 1971. Clock mutants of *Drosophila melanogaster*. *Proc. Nat. Acad. Sci. USA*, **68**; 2112–2116.

Kopeć S. 1917. Experiments on metamorphosis of insects. *Bull. int. Acad. Sci. Cracovie* (B), 57–60.

Kopeć S. 1922. Studies on the necessity of the brain for the inception of insect metamorphosis. *Biol. Bull.*, **42**; 322–342.

Kroeger H. 1971. Insect hormones and gene activity. *Endocrinol. Exp.*, **5**; 108–113.

Maddrell S. H. P. 1974. Neurosecretion. [In:] *Insect Neurobiology*, J. H. Treherne (ed.), pp. 307–357. North-Holland, American Elsevier, Amsterdam, Oxford, New York.

Muszyńska-Pytel M. 1987. Allatotropic activity of the median neurosecretory cells of *Galleria mellonella* (*Lepidoptera*) larval brain. *Experientia*, **43**: 908–909.

Novák V. J. A. 1966. *Insect Hormones*. Methuen and Co. Ltd., London.

Novák V. J. A. 1971. The research into insect endocrines. [In:] *Insect Endocrines*, pp. 11–16. Academia, Praha.

Sláma K., Romaňuk M., Šorm F. 1974. *Insect Hormones and Bioanalogues*. Springer-Verlag, Wien-New York.

Steel C. G. H. 1973. Humoral regulation of the cerebral neurosecretory system of *Rhodnius prolixus* (Stal.) during growth and moulting. *J. Exp. Biol.*, **58**; 117–187.

Williams C. M. 1956. The juvenile hormone of insects. *Nature*, **178**; 212–213.

Wigglesworth V. B. 1954. *The Physiology of Insect Metamorphosis*. Cambridge University Press, Oxford.

Wigglesworth V. B. 1964. Homeostasis in insect growth. *Symp. Soc. Exp. Biol.*, 18; 265–281.

Wigglesworth V. B. 1970. *Insect Hormones*. Oliver and Boyd, Edinburgh.

Chapter 2

The insect endocrine system

2.1. General characteristics of the insect endocrine system

The endocrine system of insects is greatly differentiated as regards its morphology, and all its modifications cannot be discussed here. In investigations of the anatomy of this system the technique of intravital staining with methylene blue is very helpful. Methods of investigating the distribution of the components of the insect endocrine system have been described earlier in the excellent papers of Chauthani and Callahan (1966), Edwards (1968) and Finlayson and Osborne (1970).

The endocrine system comprises the following structures: the brain, the *corpus allatum* (*CA*), the *corpus cardiacum* (*CC*), the prothoracic glands (PG), the frontal ganglion, the subesophageal ganglion, and the thoracic and abdominal ganglia. The location of these endocrine centres is shown in Figure 2.1. The relative proportions of the structures are not accurately represented in the diagram. The spatial distribution of these organs in the cephalic segment is illustrated in Plate 1.* It is worth noting that these organs are continually bathed in haemolymph flowing from the aorta, which opens into the cephalic segment. This facilitates the rapid diffusion of the hormones produced throughout the body.

2.2. The insect neurosecretory system

2.2.1. Neurosecretory cells of the brain

The insect brain plays a dominant role in the coordination of interactions between the external environment and the endocrine system. It consists of three basic parts: the *protocerebrum*, the *deutocerebrum* and the *tritocerebrum* (Jawłowski, 1960; Gieryng, 1965). Neurosecretory cells producing various kinds of neurohormones may be present in any of these parts. The presence of neuro-

* Plates are inserted at the end of the book.

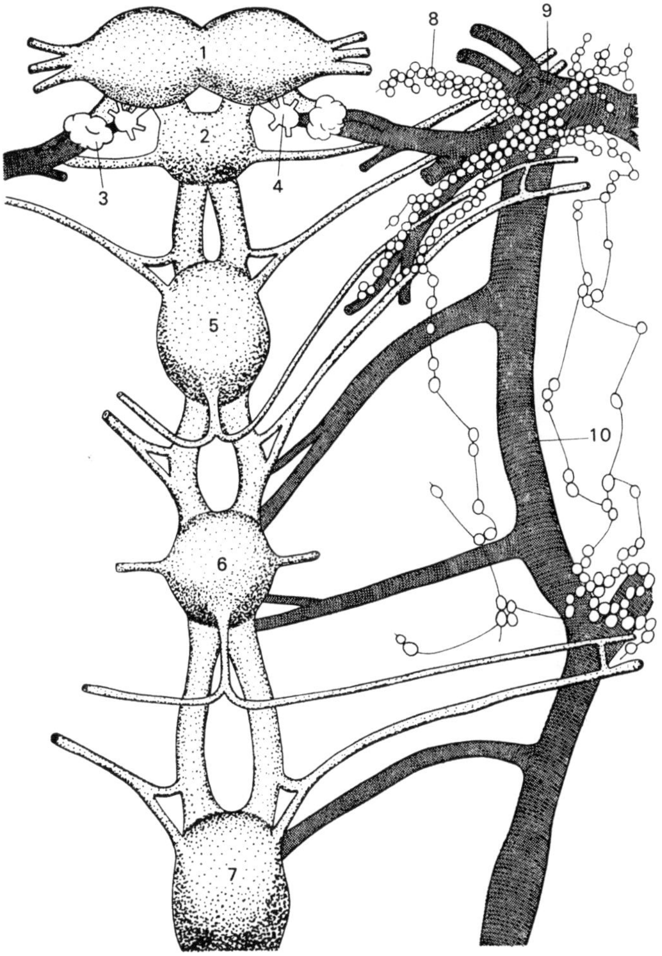

Figure 2.1. Diagram showing the distribution of the parts of the endocrine system of *Hyalophora cecropia* pupa. After Herman and Gilbert (1966).
1—brain, 2—suboesophageal ganglion, 3—*corpus allatum*, 4—*corpus cardiacum*, 5—anterior thoracic ganglion, 6—medial thoracic ganglion, 7—posterior thoracic ganglion, 8—prothoracic ganglion, 9—spiracle, 10—trachea.

secretory cells was demonstrated for the first time in histological investigations of the bee brain (Weyer, 1935), and since then they have been found in all insect orders that have been examined. The *Lepidoptera* have been the most intensively studied group in this respect. It soon appeared that neurosecretory cells are present in the brains of larvae, pupae and adults. The distribution of these cells differs among insect species, though a feature characteristic of all insects studied to date seems to be the four main groups of neurosecretory cells found in the *protocerebrum* (Fig. 2.2). Two of them lie in the *pars intercerebralis*, and there are two groups of lateral cells. The number of cells differs widely in the insects that have been studied. For instance, in both medial groups of cells in the brain

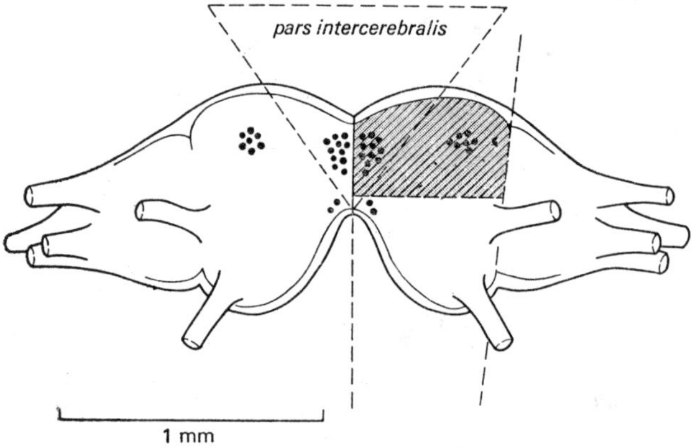

Figure 2.2. Structure of the brain of *Hyalophora cecropia* pupa. Hatched part represents endocrinally active region of brain, present in both hemispheres. Dark spots indicate distribution and number of medial (22), lateral (14) and posterior (4) neurosecretory cells in brain. After Williams (1969).

of *Hyalophora cecropia* pupae there are 22 cells (11 on each side); each lateral group consists of seven cells, and there are four in the posterior part of the brain, making a total of 40 neurosecretory cells (Herman and Gilbert, 1965; Williams, 1969). Seven morphologically different types of cells can be distinguished, and these will be discussed later in this chapter.

In the *protocerebrum* of the cricket (*Acheta domesticus*) there are as many as 900 neurosecretory cells, which may be classified into six main types (Geldiay and Edwards, 1973). In the brains of larvae and pupae of *Manduca sexta* five different groups of neurosecretory cells were detected, their position changing during the course of metamorphosis (Nijhout, 1975).

It was mainly through the availability of the cobalt method (see Chapter 3) that it became possible to detect the relations between the different neurosecretory groups in the brain and other organs of the insect (Fig. 2.3). As already mentioned, there are two groups of medial cells in the *pars intercerebralis*, their axons forming the *nervi corporis cardiaci I* (*NCC I*) which intersect in the *protocerebrum*. Further,

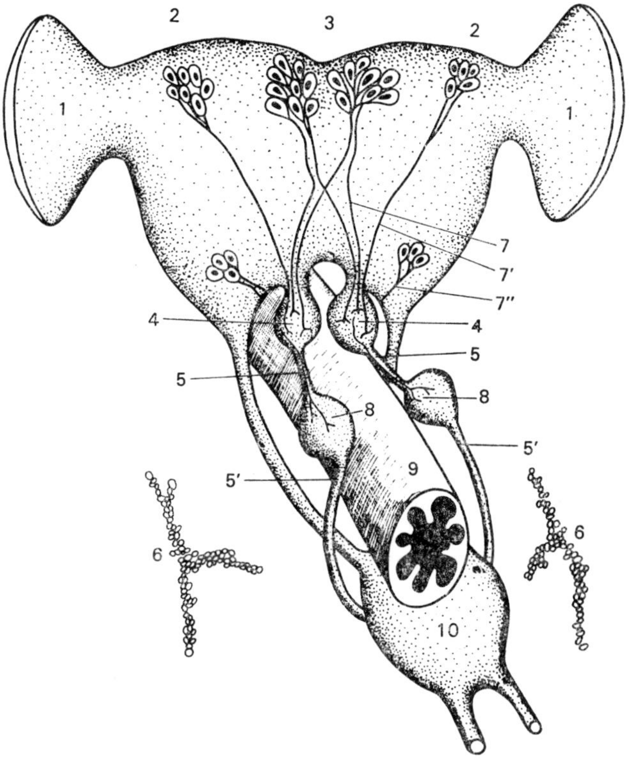

Figure 2.3. Connections between brain neurosecretory cells of insects.
1—optic lobe, 2—lateral neurosecretory cells, 3—medial neurosecretory cells, 4—*corpus cardiacum*, 5, 5'—*nervus corporis allati I* and *II*, 6—prothoracic gland, 7, 7', 7''—*nervus corporis cardiaci I*, *II* and *III*, 8—*corpus allatum*, 9—oesophagus, 10—suboesophageal ganglion.

together with axons of the lateral neurosecretory cells (*nervi corporis cardiaci II*—*NCC II*), they form nerve fibres (*NCC I*+*NCC II*) leading to the *corpora cardiaca*. The axons of the neurosecretory cells in the *tritocerebrum* form the *nervi corporis cardiaci III* (*NCC III*). The *corpora cardiaca* are joined to the *corpora allata* by means of the *nervi corporis allati I* (*NCA I*). As demonstrated in the investigations of Nijhout (1975) on larvae of *Manduca sexta*, these nerves contain both *NCC I* and *NCC II* fibres. The *corpora allata* are further connected with the suboesophageal ganglion by means of the *nervi corporis allati II* (*NCA II*).

Examination under the electron microscope showed that the *NCA II* of *Acheta domesticus* (Weber and Gaude, 1971) and the cockroach *Periplaneta americana* (Pipa and Novak, 1979) each consists of nine axons 3–4 μm in diameter, and contain the neurohaemal organ.

The neurohormones flow along these pathways and are released into the haemolymph through special neurohaemal organs. That this is so was confirmed by an experiment performed with *Leucophaea maderae*, involving unilateral section of the *NCC*. This treatment causes the accumulation of the neurosecretion in the sectioned nerve leading from the neurosecretory cells (Scharrer, 1945). Further confirmation was provided by electron microscope studies of the brain and *NCC* (Plates 2 and 3). Apart from axons of the neurosecretory cells, motor-neuron axons are also present in the *nervi corporis cardiaci*; their diameter is several times smaller than that of the neurosecretory axons (Plate 3).

2.2.2. Neurosecretory cells of other parts of the nervous system

The early part of this chapter gave an account of the detection of neurosecretory cells in the insect brain, and of the numerous studies of their morphology and functions. Similar cells may be found in other parts of the nervous system. Beattie (1971) demonstrated the existence of such cells in the optic lobes of the brain of *Periplaneta americana*. About 120 neurosecretory cells are also present in the region between the *medulla terminalis* and *lamina ganglionaris*. Their detection was relatively difficult since they do not stain with the commonly used dye aldehyde fuchsin, although they can be stained by the azane method. Their function is not yet known. However, it seems that they may act as a first link between the external environment and the endocrine system. Large amounts of tryptophan found in the vicinity of these cells, suggest that they belong to the insect's aminergic system which, according to Muszyńska-Pytel and Cymborowski (1978), may participate in the regulation of the locomotor activity of the insect.

Neurosecretory cells have also been detected in the frontal ganglion of *Manduca sexta* (Bell *et al.*, 1974), and have been examined in detail under the electron microscope (Borg *et al.*, 1973). Their function is not precisely known. It is supposed that they participate in processes such as the regulation of the water balance of the insect, the control of oocyte development and vitellogenesis (Hertel *et al.*, 1978).

There are also neurosecretory cells in the suboesophageal ganglion, but their function remains unclear. The suggestions of Harker (1960) pointed to their participation in the neurohormonal regulation of the locomotor activity rhythm in *Periplaneta americana*, implying that the biological clock would be localized in them. Further investigations did not, however, confirm these suppositions. It has been demonstrated, on the other hand, that the neurosecretory cells of the suboesophageal ganglion of the silkworm are the source of the diapause hormone (Fukuda, 1963). This hormone was isolated and purified by Isobe *et al.* (1976). It would seem, however, that with the exception of several species in which the suboesophageal ganglion actually takes part in the regulation of diapause, in most of the groups studied to date this has not been confirmed. Thus, the neurosecretory cells of this ganglion require further study.

Investigations on the structure and function of the neurosecretory cells of the thoracic and abdominal ganglia are also very scarce. Their presumed function will be discussed in subsequent parts of this book.

2.2.3. Types of neurosecretory cells

By the use of histochemical methods (see Chapter 3) various types of neurosecretory cells have been distinguished in the insect brain, to which various physiological functions are attributed. In his studies on *Oncopeltus fasciatus*, Johansson (1958) distinguished four types of neurosecretory cells in the insect brain and denoted them by successive letters of the alphabet. In the *pars intercerebralis* the most numerous are the type *A* cells, which stain purple or dark blue with aldehyde fuchsin. These cells are clearly visible *in vivo* owing to their opalescence (see Chapter 3). They usually contain large amounts of secretory granules distributed around the cell nucleus. In the brain of the cricket (*Acheta domesticus*) these cells usually lie close to the upper surface of both hemispheres (Cymborowski, 1970; Geldiay and Edwards, 1973). As in the case of *Oncopeltus fasciatus* they contain much neurosecretion (Plate 4).

Type *B* cells stain green or blue-green with aldehyde fuchsin and red in Gomori's method. They are somewhat smaller than type *A* cells and do not contain neurosecretory granules (Plate 4).

Type *C* cells lie in the same layer of the brain as the type *B* cells, but are much larger and contain neurosecretory granules sparsely distributed around the nucleus (Plate 4). Aldehyde fuchsin stains them purple-red. Cells of type *D* are relatively difficult to distinguish. In size they hardly differ from those of type *C*, but they stain differently with aldehyde fuchsin, taking on a pale red hue.

Some investigators additionally distinguish within each cell type several subtypes of neurosecretory cells, denoting them for instance A_1, A_2, etc. (Kind, 1978). It seems, however, that these subtypes may in reality be cells of the same type in various phases of synthesis and release of neurosecretion, differences in their affinity for the same stain being determined by the phase of neurosecretion at which the cells are stained.

It may be said, in general, that four to seven different types of neurosecretory cells can be distinguished in the nervous system of insects. This suggests that they may produce various neurohormones. Rowell (1976) considers that only three types of neurosecretory cells should be distinguished (types *A*, *B* and *C*), since these show distinct differences and, as demonstrated by Raabe and Monjo (1970), produce three different types of protein. This view seems acceptable until precise methods of neurosecretory cell classification are developed. The distinguishing of increasing numbers of classes of neurosecretory cells makes comparison of the results of different authors impossible without studying details of the techniques used.

Neurosecretory cells are also readily distinguishable from others when examined under the electron microscope because of the presence of large numbers of elementary secretory granules in them (Plate 5). Here the size of these granules serves as the basis for classification. For instance, in the brain of *Acheta domesticus*, the neurosecretory cells of type *A* (type I on the nomenclature of Geldiay and Edwards, 1973) contain granules 1500–3000 μm in diameter, and the cells of types *B* and *D* contain granules 600–1600 μm in diameter. In the brain of *Galleria mellonella* Wartoń and Dutkowski (1978) distinguish cells of the aminergic system containing granules 60–80 μm in diameter. It would seem, however, that in view of wide differences in the dimensions of neurosecretory granules, their size cannot serve as a basis for the classification of neurosecretory cells. The size of these granules may be dependent on the stage of synthesis or release of neurosecretion.

2.2.4. Cycle of synthesis and release of neurosecretion

By staining with aldehyde fuchsin, it is possible to follow the cycle of accumulation and release of neurosecretion. In investigations on *Acheta domesticus* (Cymborowski, 1970) it was found that cells of type *A* exhibit a noticeable circadian rhythm of accumulation and release of neurosecretory material (Fig. 2.4). In the initial stage of accumulation, the neurosecretory cells have a large nucleus with pronounced chromatin staining. Within the cytoplasm, droplets of neurosecretion of a rather irregular shape are visible (Fig. 2.4 A). In the region where secretion is formed, the cytoplasm is more dense than at the nucleus and at the cell periphery. The next step is characterized by the presence of single or grouped droplets of neurosecretion of a distinctly spherical shape (Fig. 2.4 B).

As shown by electron microscope studies, the granules become spherical owing to condensation of the neurosecretory material in the Golgi apparatus.

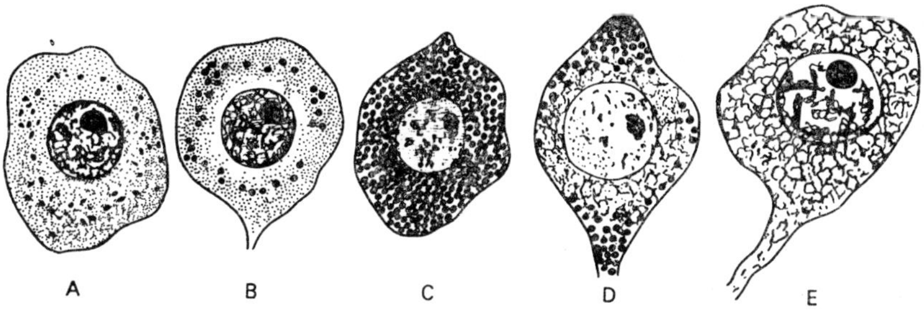

Figure 2.4. Cycle of accumulation and release of neurosecretory material by type *A* neurosecretory cells in brain of *Acheta domesticus*. Description of steps in text.

When they emerge from this structure they are surrounded by a delicate membrane. It is supposed that this membrane originates from the Golgi complex, and probably contains phospholipoproteins (Bern, 1962). At this stage the chromatin continues to stain strongly within the nucleus. A large nucleolus also appears. At a later stage (Fig. 2.4 C) a large accumulation of neurosecretory granules is observed. In the centre of the cytoplasm these granules coalesce into a single dark spot. The cell nucleus is compressed, appearing smaller than in previous stages. Within the nucleus, weakly staining chromatin is visible. The nucleolus has an indistinct outline and stains very poorly. The next step is the release of the neurosecretion from the cell. The granules accumulate at the base of the axon (Fig. 2.4 D). After release of the secretion from the cell (Fig. 2.4 E) numerous vacuoles may be seen in the cytoplasm; these are perhaps the remainder of the neurosecretory granules. The nucleus is large, with strongly staining chromatin. Noteworthy is the large, clearly outlined nucleolus.

The cycle of accumulation and release of the neurosecretory cells of the brain can also be observed under *in vitro* conditions. For instance the brain of *Leucophea maderae* shows the greatest accumulation of neurosecretory material in the medial neurosecretory cells on the third day after placing the tissue in culture (Marks *et al.*, 1973). The presence of secretory granules in the axons of these cells is evidence that the transport of neurosecretory material is not inhibited *in vitro*.

The picture presented here of the cycle of accumulation and release of neurosecretory material gives an idea of the difficulties that may be encountered in the classification of neurosecretory cells. Classification of the cells as this or that type solely on the basis of the presence or absence of neurosecretory granules may be highly misleading. Similarly, when only the degree of staining produced by the neurosecretory material in the cell is observed, one cannot draw reliable conclusions about the cell's synthetic activity or the dynamics of release of the secretion. From a theoretical point of view there may be four variants of this process (Fig. 2.5). Thus, a cell containing relatively little stainable neurosecretion may be inactive, but it may also be highly active, since it may synthesize the neurohormone intensively and release it immediately. The same may be true of cells containing a large amount of stainable secretion. They may increase the rate of synthesis and release of the neurohormone without changing the amount of hormone present in the cell.

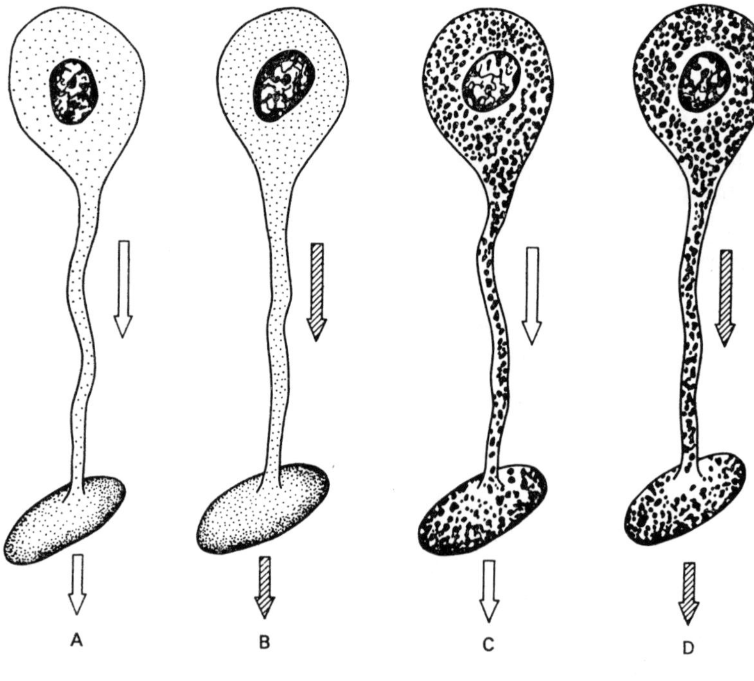

Figure 2.5. Four possible interpretations of histological data on neurosecretory cells containing various amounts of accumulated neurosecretion.

Hatched arrows indicate intensity of synthesis, transport and release of neurosecretion. A—cell contains little stained neurosecretion; it synthesizes and releases little material. B—the same cell, in spite of low content of neurosecretory material may synthesize and release it intensively. C—cell contains much stainable neurosecretion, but may synthesize and release little, D—the same cell may synthesize and release much secretory material.

For the reasons presented here, autoradiographic methods are used increasingly often in investigations on neurosecretory cells (Highnam and Mordue, 1987; Girardie and Girardie, 1972; Cymborowski 1973). In autoradiographic studies on the cricket (Cymborowski, 1973) and the wax moth (Cymborowski and Kryspin, 1975), it was demonstrated that during the period of maximal accumulation of neurosecretion (stained with aldehyde fuchsin) the rate of synthesis is very low.

Can the activity of neurosecretory cells in producing neurohormones be studied by means of autoradiographic methods ? The problem is not as simple as it seems. Many authors have reported that the contents of the neurosecretory granules visible in the electron microscope are not homogeneous. They are membrane-bounded bodies containing biologically active polypeptides and carrier proteins (Carlisle *et al.*, 1955). According to Scharrer and Scharrer (1954), it is not the hormone itself but only the carrier substance that stains when Gomori's technique is used. The carrier substance seems to be a lipid, because it is washed out when organic solvents are used. Mahon and Nair (1975) suggest that the biologically active part of the neurosecretion may be stained with alcian blue (see Chapter 3). They submit rather convincing evidence to confirm this supposition.

In answer to our question, we can say that the use of autoradiographic methods does not ensure that it is the rate of synthesis of the neurohormone itself that is under investigation, because the process being observed concerns only the carrier protein which is being intensively synthesized at the moment of observation. Of course, it seems logical to presume that both the carrier protein and the hormone are produced simultaneously in the cell, though, at present we have no proof of this.

2.2.5. Mechanisms of release of neurosecretion from the cells

So far we have discussed only processes leading to the accumulation of neurosecretion in the cells and its translocation along the axons to sites of release. At these sites, referred to as neurohaemal organs, transient accumulation of the neurosecretion takes place before its release into the haemolymph.

2.2.5.1. *Corpora cardiaca (CC)*

In the majority of insect species studied to date the most important neurohaemal organ is the *corpus cardiacum* (*CC*), which is usually a paired structure (the *corpora cardiaca*). Their position with respect to the brain and the *corpora allata* differs among species (Fig. 2.6). In extreme cases there is a ring gland containing the *corpus cardiacum*, as observed in *Calliphora erythrocephala* (Fig. 2.6) and *Drosophila melanogaster* (Fig. 2.7).

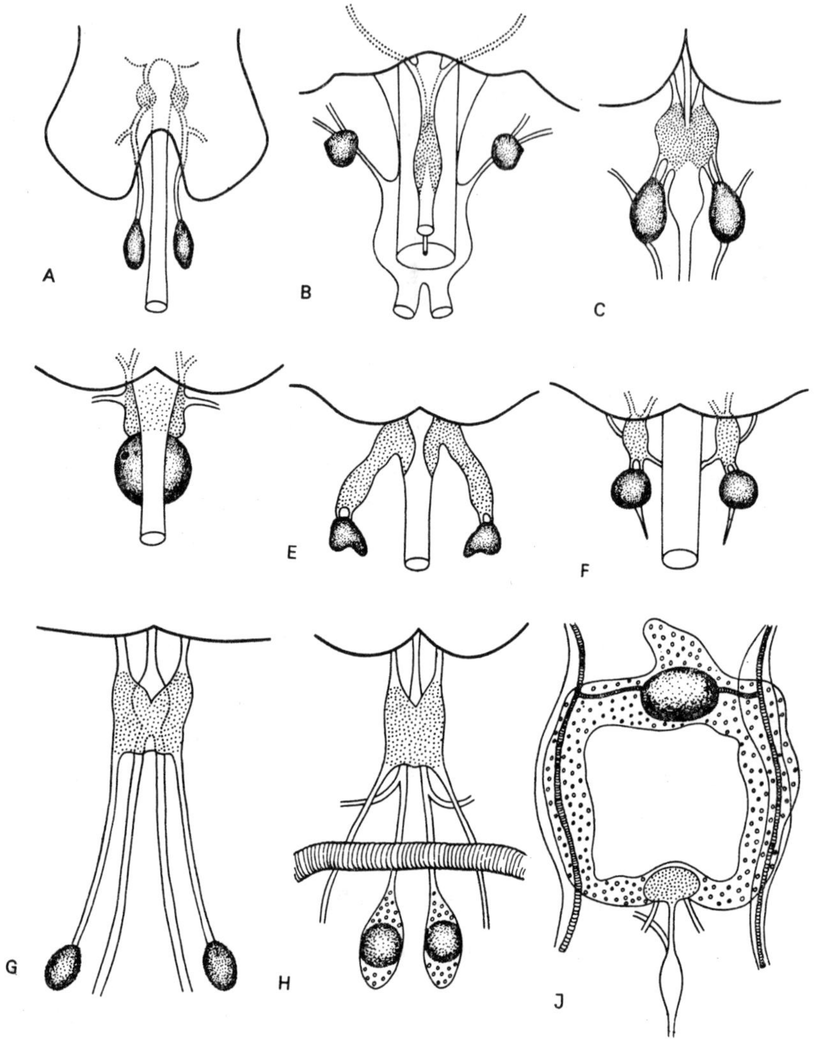

Figure 2.6. Retrocerebral system of various insect groups. After Novák (1975), with permission. Stippled areas—*corpora cardiaca*, black structures—*corpora allata*; small circles denote pericardial organ. A—*Japys* sp., B—*Ephemera vulgata*, C—*Blatta orientalis*, D—*Pyrrhocoris apterus*, E—*Hydrous piceus*, F—*Sphinx ligustri*, G—*Stalis luteria*, H—*Culex* sp., I—*Calliphora erythrocephala*.

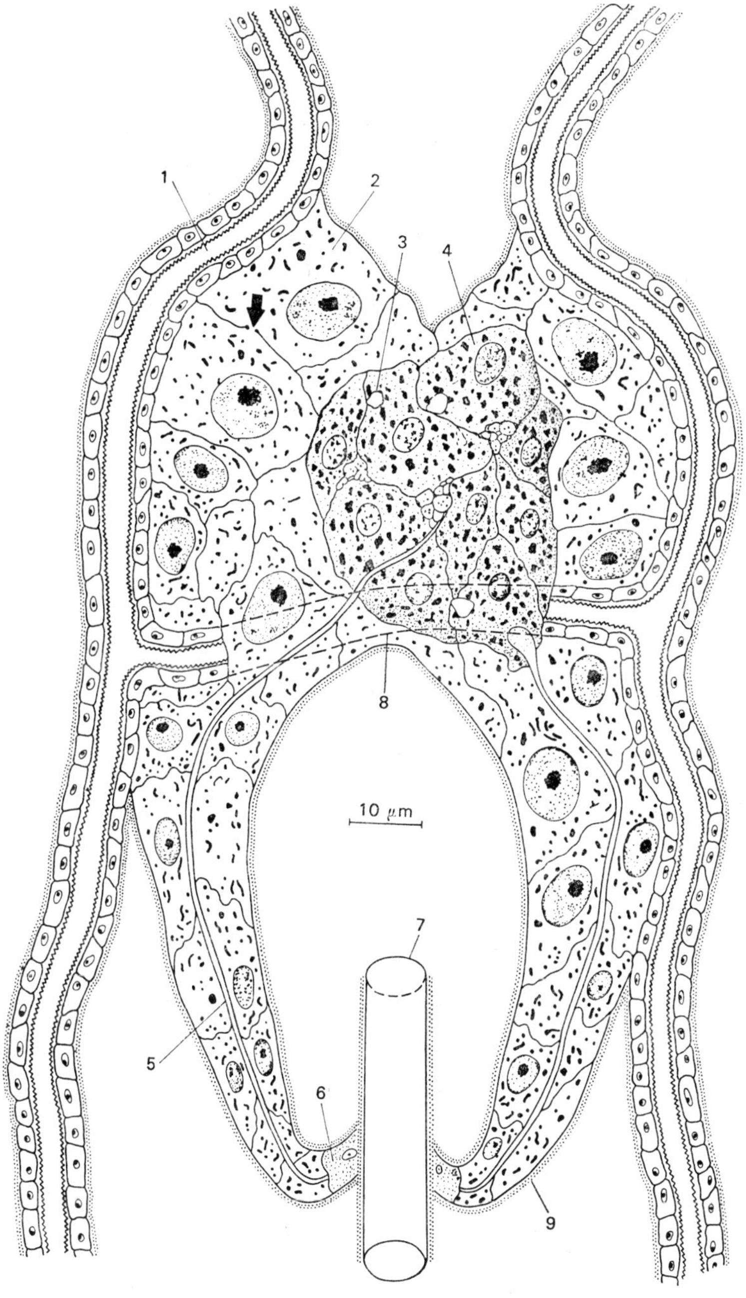

Figure 2.7. Structure of ring ganglion in third larval stage of *Drosophila melanogaster*. After King *et al.* (1966).

Black structures in cytoplasm of *corpus allatum* (4) and prothoracic gland (2) denote mitochondria. Dark spots in *corpus allatum* and dots in *corpus cardiacum* (6) correspond to neurosecretory granules. 1—left lateral trachea, 3—space left by released neurosecretory material, 5—axon leaving cell in *corpus cardiacum* and leading to *corpus allatum*, 7—aorta, 8—transverse trachea, 9—cell-free tunica.

Arrow indicates gaps in membranes of neighbouring prothoracic gland cells.

Many axons of neurosecretory cells end in these organs (Fig. 2.8), and examination under the electron microscope reveals a considerable accumulation of neurosecretory granules in them (Plate 6). Apart from the nerve endings, there are also secretory cells in the *corpora cardiaca*, specific to this organ (Scharrer, 1963). Thus, as well as the accumulation and release of neurosecretion the *CC* produce their own hormones, of which we shall speak in later parts of this book.

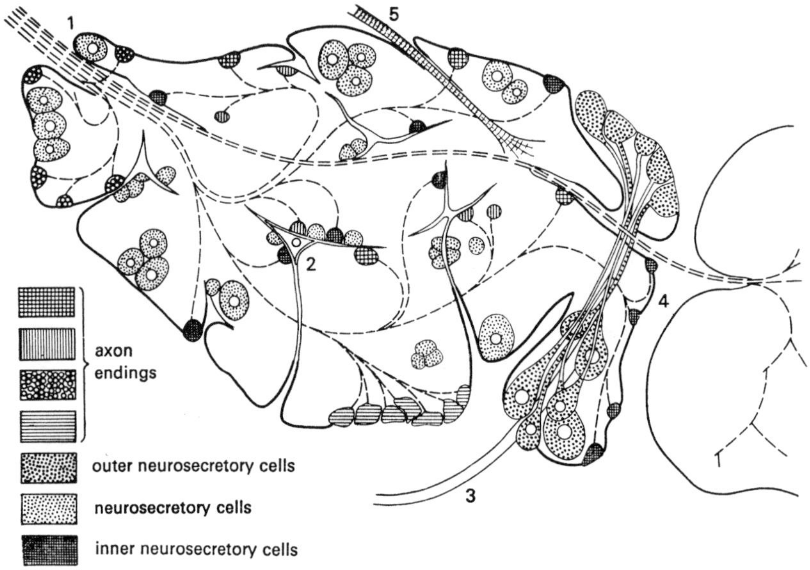

Figure 2.8. Structure of *corpus cardiacum* of adult Colorado potato beetle. After Schooneveld (1970).

Broken lines mark courses of axons of neurosecretory cells. 1—*nervi corporis cardiaci*, 2—haemocytes, 3—*nervus cardiaco-suboesophagealis*, 4—*nervus corporis allati*, 5—muscles.

What is the mechanism whereby neurosecretion is released from the nerve endings into the *corpora cardiaca*? This problem has been considered by many authors, but there is still no uniformity of opinion in this respect (Przełęcka and Cymborowski, 1970). On the basis of far-reaching histochemical investigations on type *A* neurosecretory cells in the brain of *Oncopeltus fasciatus*, Schreiner (1966) reaches the conclusion that there are, in principle, three different possibilities for the release of neurosecretion (Fig. 2.9). From a theoretical point of view, neurohormones may be present in the neurosecretory granules in one of three forms: as monomeric or polymeric hormones, or else bound to carrier proteins. If we assume that hormones occur in monomeric form in the granules, it is possible that at the moment of release (Fig. 2.9 A) the granule membrane becomes permeable. If, however, the hormone is polymerized in the Golgi apparatus after synthesis leading to the formation of a large molecule, then many such molecules may be present in the granules (Fig. 2.9 B). Thus, depolymerization must occur at the moment of release. The hormones may be released in the form of monomers to the neurohaemal oragn as described above. In both cases (Fig. 2.9A and B) there should be "empty" vesicles in the cytoplasm of the cells that have released neurosecretory material, and these can actually be observed in electron micrographs of *CC* (see Plate 6).

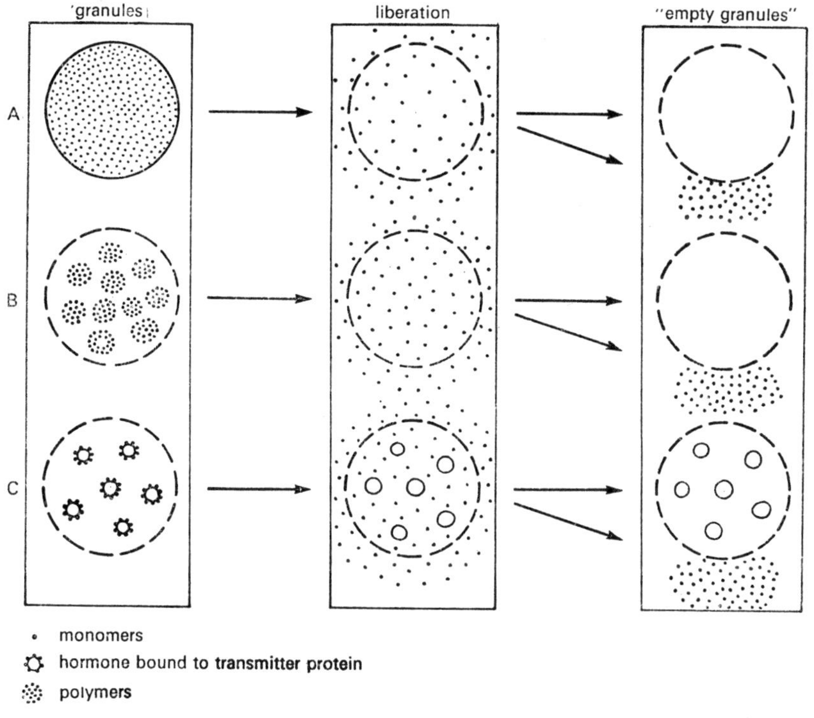

Figure 2.9. Three possibilities of hormone release from nerve endings. After Schreiner (1966). A—granule contains hormones in monomeric form (dots), B—granule contains hormones in polymeric form, C—granule contains hormones bound to carrier proteins. Further details in text.

There is also the possibility that the neurohormones are not present in polymerized form in the granules. If this is the case, one can imagine that the carrier proteins are indispensable, binding to the monomeric forms of the hormone in order to prevent "leakage" from the granules before the appropriate moment.

Before release, the hormone would separate from the carrier protein, which would remain in the vesicle after release of the hormone.

Investigations on the stick insect (*Carausius morosus*) demonstrated that the hormone passes directly from the axonal endings of the neurosecretory cells into the intercellular spaces connected to the vascular system (Smith and Smith, 1966). The same authors also demonstrated that the secretory product of the specific neurosecretory cells of the *corpora cardiaca* passes in the same way into the haemolymph (see Plate 6).

2.2.5.2. The perisympathetic system of insects

The detailed studies of Raabe (1965, 1967) and her co-workers led to the detection of neurohaemal organs metamerically distributed along the entire length of the nerve chain (Fig. 2.10). These organs are the site of accumulation and release of the secretory material produced by the neurosecretory cells of the individual nerve ganglia, both in the thorax and the abdomen. They appear as distentions of the visceral nerves, and are called perisympathetic organs because they are connected with the sympathetic nerve system.

Histologically, the perisympathetic organs do not differ from the *corpora cardiaca.* They contain numerous nerve endings of neurosecretory cells, with granules of various sizes. Brady and Maddrell (1967) found that in the perisympathetic system of *Carausius morosus* electron-dense and -clear granules of diameter 1100–2500 μm are present; in the locust (*Schistocerca gregaria*) the granule diameter is 400–800 μm.

In addition to the endings of neurosecretory cells, the perisympathetic system has its own secretory cells, mainly of type *C*, but also (rarely) of type *A*, as was found after examining 24 species of *Coleoptera* (Provansal and Grillot, 1972), eight species of *Diptera* (Grillot, 1977) and seven of *Heteroptera* (Baudry-Partiaoglou, 1978).

Using the cobalt technique of cell staining (see Chapter 3) Ali and Pipa (1978) studied the nervous communications between the perisympathetic system and the neurosecretory cells of the third abdominal ganglion in *Periplaneta americana.*

2.2.5.3. Other neurohaemal organs of insects

In addition to the neurohaemal organs described above, Geldiay and Edwards (1973) detected sites fulfilling a similar role in the brain of the cricket (*Acheta domesticus*). In the ventral part of the brain and immediately under its surface

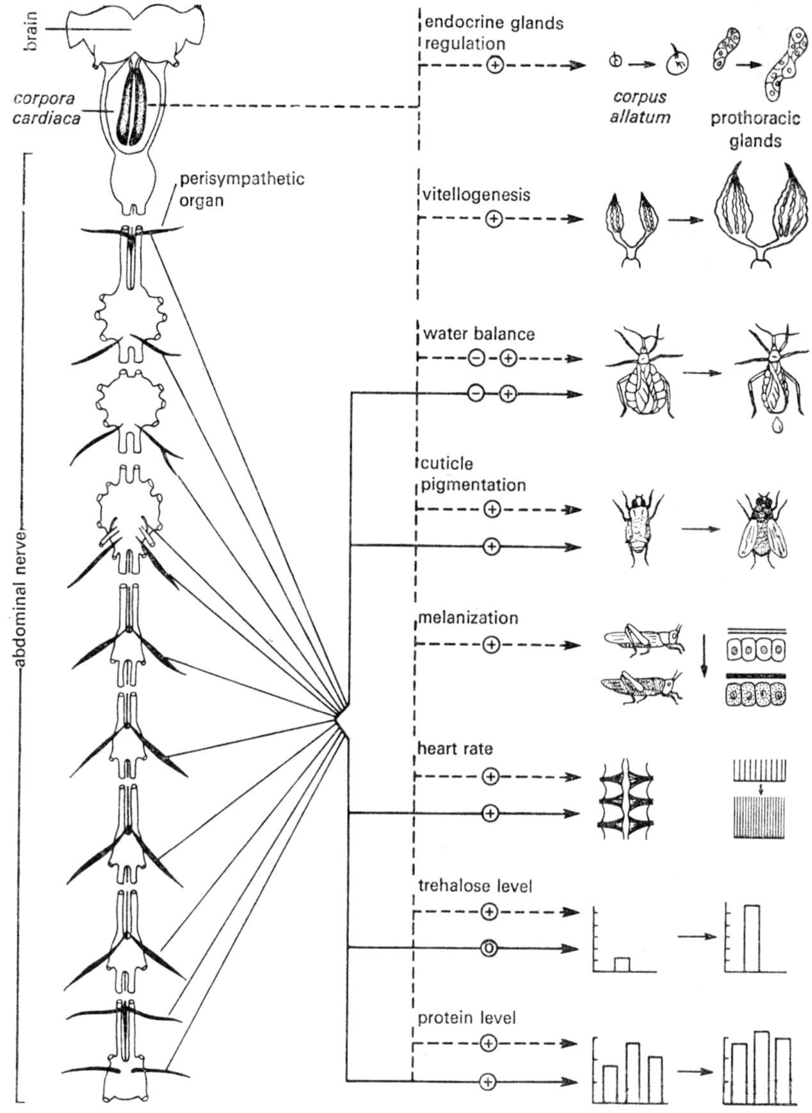

Figure 2.10. Distribution of insect neurohaemal organs (perisympathetic system) in relation to their physiological function. Plus denotes stimulation, minus inhibition of particular organs or processes. After Raabe *et al.* (1974) with permission.

there is an area in which type *B* and *D* neurosecretory cell nerve endings are present, containing granules 700–1600 μm in diameter. The function of these neurohaemal areas has so far not been elucidated.

Recently, a number of papers have appeared indicating that the *corpora allata* can also serve as neurohaemal organs (Tombes and Smith, 1970; Gibbs and Riddiford, 1977). Agui *et al.* (1979) showed that the *CA* of *Manduca sexta* contain two to nine times more prothoracicotropic hormone (PTTH) than the *corpora cardiaca*, which were considered to be the only neurohaemal organ releasing this hormone. Further research is necessary to elucidate the mechanism of PTTH release by the *CA*.

2.3. *Corpora allata (CA)*

Intensive investigations of the *corpora allata* as endocrine glands began with Wigglesworth's (1935) detection of their inhibitory effect on metamorphosis and their stimulating influence on the development of the ovaries. It is now known that these effects result from the production of juvenile hormone and its release into the haemolymph.

The morphology of the *CA* and their situation in relation to the brain and the *corpora cardiaca* have already been described (see Fig. 2.6). Histological investigations have shown that wide structural differences exist among insect species (Fig. 2.11). The simplest structures consists of a few large parenchymal cells with polymorphic nuclei (Fig. 2.11 A). The cells are usually connected to each other by connective tissue. In other insect groups the *CA* usually show a more compact structure, and the parenchymal cells are uniformly distributed throughout the entire organ (Fig. 2.11 B). Further differences in histological structure concern the size of the parenchymal cells of the *CA* and their distribution around the periphery of the organ (Fig. 2.11 C).

The *corpora allata* are surrounded by a relatively thick layer of connective tissue (neurilemma) with numerous invaginations into the surface (Fig. 2.12). The parenchymal cells have rather irregular shapes, as do their nuclei. The cytoplasm contains numerous vesicles and various secretions (Waku and Gilbert, 1964; Morohoshi *et al.* 1976).

Many authors consider the size of the *CA* to be an indicator of their activity (Schooneveld *et al.*, 1977). Certain characteristic features of their structure have also been shown to be associated with activity of the glands. Such features include the presence of a large number of mitochondria and ribosomes in the cytoplasm,

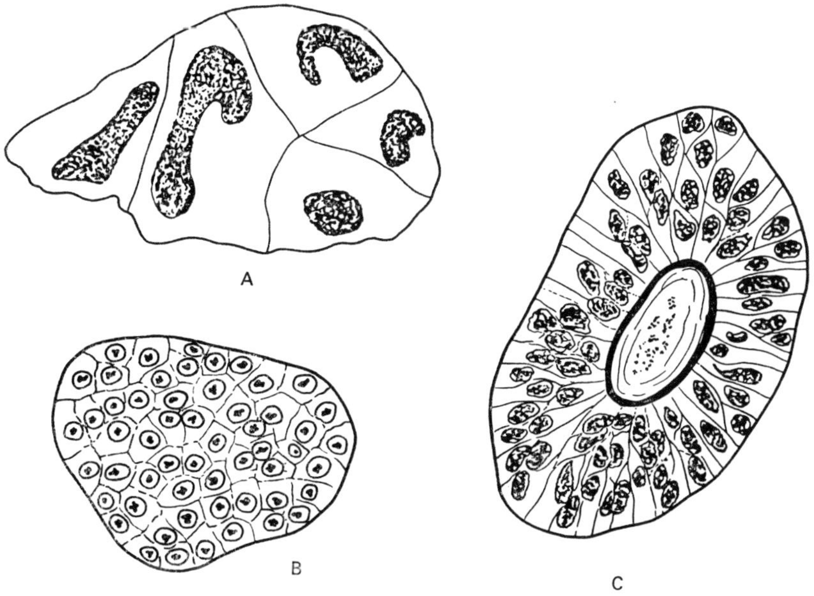

Figure 2.11. Diagrammatic representation of various types of *corpus allatum*. After Gorbman and Bern (1966).

A—*corpus allatum* of *Pieris brassicae*, consisting of several cells with polymorphic nuclei, B—*corpus allatum* of *Pentaloma rufipes*, consisting of a large number of small cells with poorly outlined membranes, almost a syncytium, C—*corpus allatum* of *Bacillus rossii* with space in centre into which the hormone is secreted.

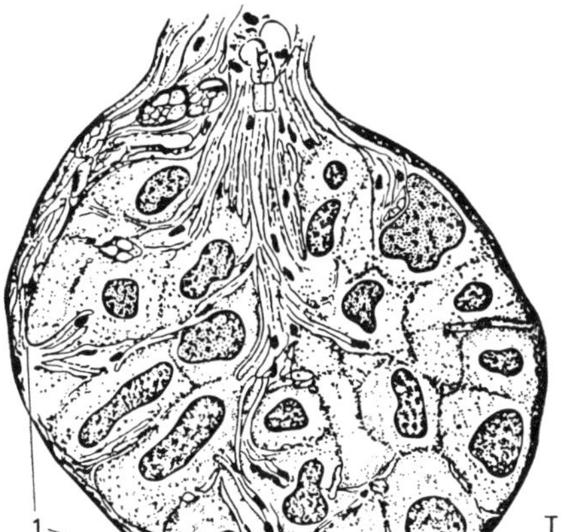

Figure 2.12. Structure of *corpus allatum* in *Hyloicus* pupa showing internal structure and innervation. Note large cells with polymorphic nuclei. 1—axon endings of *nervus corporis allati*. After Gorbman and Bern (1966).

and large irregularly shaped nuclei with nucleoli. Scharrer (1971) describes electron-dense structures in the region of the Golgi apparatus. Their distribution and number change with the stage of development of the insect, suggesting that they participate in the secretory function of the *CA*. Odhiambo (1966) believed that synthesis of the most important hormone, juvenile one (JH), occurs in the canaliculi of the endoplasmic reticulum and that the hormone is "enclosed" in the vesicles. During the period in which the secretion is released into the haemolymph of the insect the vesicles are dissolved in lipids, allowing the hormone to pass through the membrane surrounding the *CA* (Thomsen and Thomsen, 1970).

In the successive larval instars the volume of the *CA* increases more than two-fold; this growth occurs without any increase in the number of cells, and results solely from cell enlargement. During metamorphosis the histological structure of the *CA* indicates that its cells are inactive. This is manifested in the first place by a decrease in the volume of the cell nucleus and by distinct changes in the cytoplasm. In newly emerged imagines *CA* activity is renewed. In many species characteristic cyclic changes in the size of the *CA* are observed in the adults correlated with the cyclic development of the ovaries, and periods of feeding and oviposition (Adams, 1970). This correlation is particularly pronounced in the cockroach (Engelmann, 1970).

Changes in the size of the *CA* are also associated with the diapause in insects. A good example of this is the Colorado potato beetle, the object of many experiments conducted by de Wilde (1968). These problems will be discussed later in this book.

The differences in *CA* size among the castes of social insects should be mentioned here. Investigations have been performed particularly on termites (Lüscher, 1960) in which the castes differ widely from one another (soldiers, workers, etc.).

2.4. Prothoracic glands (PG)

2.4.1. Anatomical and microscopic structure of the prothoracic glands

The prothoracic glands are paired organs consisting of several branched groups of secretory cells producing the moulting hormone, ecdysone. They lie on either side of the nerve chain in the ventrolateral region of the prothorax. In some species (e.g. *Manduca sexta*) branches of the glands reach to the cephalic segment in the anterior part of the body and to the mesothorax in the posterior part. Their external morphology is known from the studies of Lee (1948). Various types of prothoracic glands are shown in Figure 2.13. The individual cells of the gland are joined by connective tissue strands, but the connections are rather loose so that the glands often disintegrate when removed. They are usually attached to the main tissues running along the sides of the body, and are well innervated. In the *Lepidoptera* for instance, the nerves of the suboesophageal

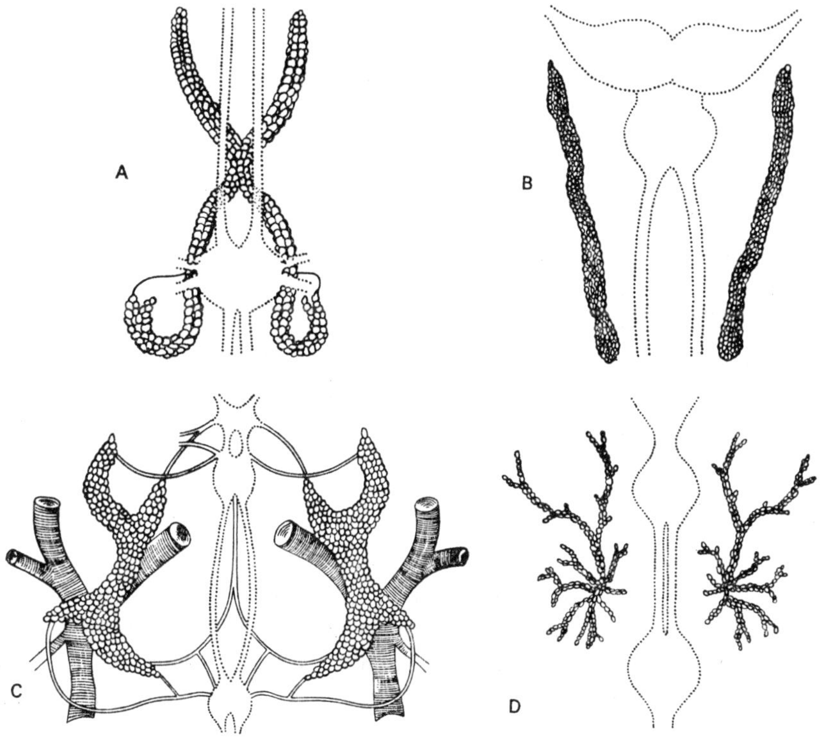

Figure 2.13. Various types of prothoracic glands. After Novák (1975) with permission. A—*Blatioptera*, B—*Hemiptera*, C—*Lepidoptera*, D—*Hymenoptera*.

ganglion and the first thoracic ganglion, and branches of the connectives joining the first and second thoracic ganglia, extend to the PG, as demonstrated by Herman and Gilbert (1966) and Singh and Sehnal (1979).

Examination under the electron microscope showed that in the axons of neurons present in the PG of *Galleria mellonella* there are neurosecretory granules 120×150 nm in size (Benedeczky *et al.*, 1980). This suggests the possibility that the function of the PG is regulated not only by bioelectric activity, but also by neurohormonal influences.

The number of cells constituting the PG varies. In *Galleria mellonella*, for instance, there are about 55 in each gland. The nuclei of these cells are enormous; they are of irregular shape and almost completely fill the cell (Plate 7). These nuclei are polyploid. The cytoplasm is in the form of a narrow strand, and increases in size only during the period of greatest secretory activity of the glands. Electron microscope studies demonstrated the presence of micropinocytotic vesicles within the cytoplasm. It is supposed that these vesicles participate in the process of secretion of ecdysone into the haemolymph. The mechanism of this process has not so far been elucidated.

2.4.2. Changes in structure and function of PG in insect development

As shown by Malá *et al.* (1974), the structure of the prothoracic glands in *Galleria mellonella* changes during the development of the insect. Criteria indicating a high secretory activity of the glands have been established. These include an increase in cell volume, changes in shape and dimensions of cell nuclei, and the pinocytotic vesicles mentioned previously.

Notwithstanding the type of metamorphosis (complete or incomplete) the size of the PG increases with successive larval moults. However, not all cells in these glands increase their dimensions synchronously. During each stage, a proportion of them seem to be inactive and to increase their activity at a later stage (Malá *et al.*, 1974). Usually, on the first day after moulting, the cytoplasm is thin and the nucleus has a smooth surface (Fig. 2.14). On the next day the cytoplasm becomes thicker and the surface of the nucleus shows folds. During the period when the new cuticle starts to separate from the old one (apolysis) PG activity, evaluated on the basis of morphological changes, reaches its maximum, and the cytoplasm is at its thickest and contains a vacuole. The nucleus surface is densely folded. On the next day the cytoplasm becomes thinner and the nuclear surface becomes smoother. Around this time the insect moults to the next instar. At the beginning of the last instar the cytoplasm is also very thin, but it forms two layers. The outer layer is very dense compared with the inner one. There are a number of vesicles on the cell surface that are not observed in the previous larval stages (Fig. 2.14). On the successive days of the last instar the cytoplasm thickens and vacuoles connected with the nuclear envelope appear. More and more vesicles now appear immediately under the cell surface, so that it becomes rough. This seems to be the period of maximum PG activity. At this time the PG

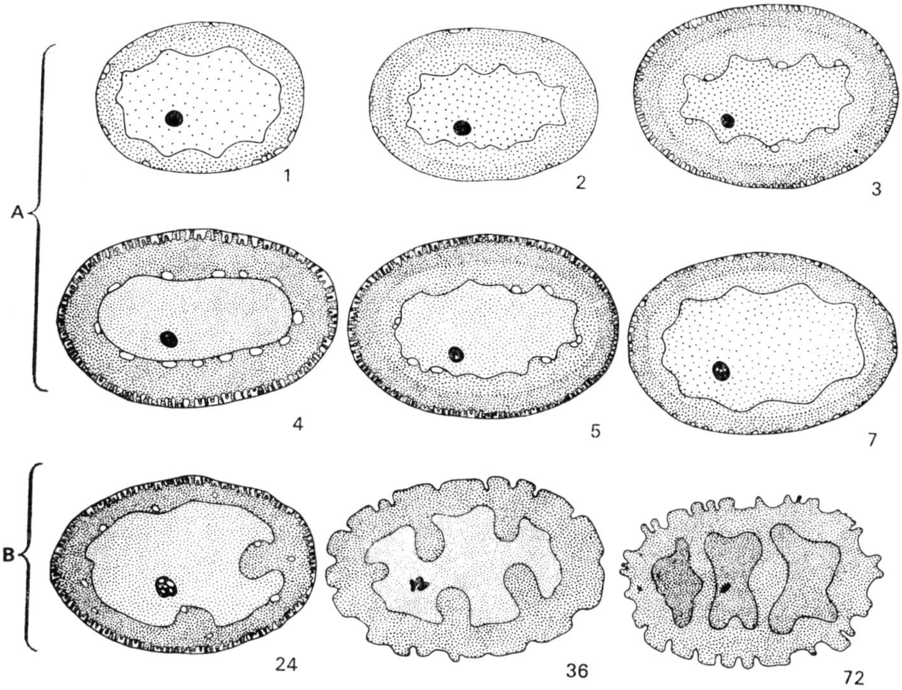

Figure 2.14. Morphological and functional changes in cells of prothoracic glands of the wax moth (*Galleria mellonella*) during the last larval stage and pupal stage. After Mala *et al.* (1974) with permission.

A—larval stage: successive numbers denote duration of last stage, B—pupal stage; numbers indicate hours elasped after pupation. Further details in text.

returns to the structure it showed during the first days of the last instar. PG cells of the newly developed pupa still show secretory activity, but soon changes appear indicating involution of the cells within the gland. The first of these changes arise in the nucleus; vacuoles appear and the nucleus then undergoes fragmentation. Soon afterwards the cell surface begins to fold, leading to separation of the cells and their complete disintegration (Fig. 2.14 B). Noteworthy is the fact that the process of cell disintegration occurs gradually, so that single cells may persist into the first days of imaginal life. It is not known, however, whether such single cells are still capable of producing and releasing ecdysone.

The structural and functional changes in the prothoracic glands in postembryonic development of *Galleria mellonella* described here are characteristic of all insects examined to date. A review of these problems is given by Sláma *et al.* (1974, pp. 49–58).

2.5. Summary

The insect endocrine system shows a high degree of morphological differentiation. It consists basically of three main endocrine centres: the neurosecretory system, the *corpora allata* and the prothoracic glands. The neurosecretory system comprises neurosecretory cells present mainly in the brain, but also in various other parts of the nervous system. Beside their normal bioelectric activity these cells produce several kinds of protein hormones and deliver them to special neurohaemal organs in which they are temporarily stored before release to the haemolymph. Apart from peptidergic cells producing protein hormones, there are aminergic cells which supply various biologically active derivatives of amino acids, such as 5-hydroxytryptamine (serotonin). The functions of both these systems are at present under intensive study.

Research in recent years has shown that, in addition to the *corpora cardiaca*, the insect nervous system contains many other neurohaemal centres. A great surprise was the finding that the *corpora allata*—the glands that produce juvenile hormone—also act as neurohaemal organs. Many discoveries have been made concerning the mechanisms of accumulation and release of the neurosecretory material in these organs. These discoveries became possible through the use of the electron microscope. Other investigations on the neurosecretory system deal with the functions of individual cells and of the chemical structure of the neurohormones produced by them. In this respect there still remains a great deal to be done. The discovery of the metamerically organized perisympathetic system, which functions as a neurohaemal organ for various neurosecretory cells in the ganglia of the ventral nerve chain, opened up a wide range of possibilities for the study of the functions of the insect neurosecretory system as a whole, and of its component parts. Further studies, particularly those involving the use of the cobalt technique to elucidate the nervous connections between the neurosecretory and the perisympathetic systems, will provide new data on the function of individual neurosecretory cells throughout the insect nervous system.

There have also been many interesting discoveries concerning the fine structure of the *corpora allata* and prothoracic glands. Detailed criteria for morphological changes in relation to functional changes in these organs have been established. Particularly noteworthy is the elucidation, at least in part, of the mechanism of release of the hormones produced both by the *corpora allata* and by the prothoracic glands. Further histochemical investigations on this subject are required.

This chapter does not discuss the role of the many other endocrine organs, such as the gonads, the fat body and the alimentary tract. We shall return to them in further chapters of this book.

2.6. References

Adams T. S. 1970. Ovarian regulation of the *corpus allatum* in the housefly, *Musca domestica*. *J. Insect Physiol.*, **16**; 349–360.

Agui N., Granger N. A., Gilbert L. J., Bollenbacher W. E. 1979. Cellular localization of the insect prothoracicotropic hormone: *In vitro* assay of a single neurosecretory cell. *Proc. Nat. Acad. Sci. USA*, **76**; 5694–5698.

Ali Z. J., Pipa R. 1978. The abdominal perisympathetic neurohemal organs of the cockroach *Periplaneta americana*: Innervation revealed by cobalt chloride diffusion. *Gen. Comp. Endocrinol.*, **36**; 396–401.

Baudry-Partiaoglou N. 1978. Anatomie et histologie des organes neurohémaux de quelques hémiptères. *Int. J. Insect Morphol. Embryol.*, **7**; 1–31.

Beattie T. M. 1971. Histology, histochemistry, and ultrastructure of neurosecretory cells in the optic lobe of the cockroach, *Periplaneta americana*. *J. Insect Physiol.*, **17**; 1843–1855.

Bell R. A., Borg T. K., Ittycheriah P. I. 1974. Neurosecretory cells in the frontal ganglion of the tobacco hornworm, *Manduca sexta*. *J. Insect Physiol.*, **20**; 669–678.

Benedeczky J., Malá J., Sehnal F. 1980. Ultrastructural study on the innervation of prothoracic glands in *Galleria mellonella*. *Gen. Comp. Endocrinol.*, **41**; 400–407.

Bern H. A. 1962. The properties of neurosecretory cells. *Gen. Comp. Endocrinol.*, Suppl. **1**; 117–132.

Borg T. K., Bell R. A., Picard D. J. 1973. Ultrastructure of neurosecretory cells in the frontal ganglion of the tobacco hornworm, *Manduca sexta* (L.). *Tissue and Cell*, **5**; 259–267.

Brady J., Maddrell S. H. P. 1967. Neurohaemal organs in the medial nervous system of insects. *Z. Zellforsch. mikrosk. Anat.* **76**; 389–404.

Carlisle D. B., Dupont-Raabe M., Knowles F. G. 1955. Recherches préliminaires relatives à la sèparation et à la comparaison des substances chromactives des Crustacées et des Insectes. *C.R. Acad. Sci. Paris*, **240**; 665.

Chauthani A. R., Callahan P. S. 1966. A dissection technique for studying internal anatomy of different stadia of *Noctuidae*. *Ann. Entomol. Soc. Amer.*, **59**; 1017–1018.

Cymborowski B. 1968. Badania nad neurohormonalną regulacją dobowego cyklu aktywności ruchowej świerszcza domowego (*Acheta domesticus* L.) [Investigations on neurohormonal regulation of diurnal motor activity cycle in the cricket (*Acheta domesticus* L.)]. Ph.D. thesis, Univ. of Warsaw.

Cymborowski B. 1970. Investigations on the neurohormonal factors controlling circadian rhythm of locomotor activity in the house cricket, *Acheta domesticus* L. II. Daily histochemical changes in the neurosecretory cells of the *pars intercerebralis* and suboesophageal ganglion. *Zool. Polon.*, **20**; 127–153.

Cymborowski B. 1973. Les variations diurnes de l'activité des cellules neurosécrétrices du cerveau du grillon (*Acheta domesticus* L.). *Acrida*, **2**; 299–306.

Cymborowski B., Kryspin I. M. 1975. Effect of reserpine on development and its neuroendocrine regulation in *Galleria mellonella*. *Gen. Pharmac.*, **6**; 209–213.

Edwards L. J. 1968. Dissection of the endocrine organs—*corpus allatum*, *corpus cardiacum* and ecdysial (prothoracic) gland—of the larva of *Heliothis zea*. *Ann. Entomol. Soc. Amer.*, **61**; 965–968.

Engelmann F. 1970. *The Physiology of Insect Reproduction*. p. 307. Pergamon Press, Oxford.

Finlayson L. H., Osborne M. P. 1970. Electrical activity of neurohormonal tissue in the stick insect, *Carausius morosus*. *J. Insect Physiol.*, **16**; 791–800.

Fukuda S. 1963. Déterminisme hormonal de la diapause chez le vers á soie. *Bull. Soc. zool. Fr.*, **88**; 151–179.

Geldiay S., Edwards J. S. 1973. The protocerebral neurosecretory system and associated cerebral neurohemal area of *Acheta domesticus*. *Z. Zellforsch. mikrosk. Anat.*, **145**; 1–22.

Gibbs D., Riddiford L. M. 1977. Prothoracicotropic hormone in *Manduca sexta*: localization by a larval assay. *J. exp. Biol.*, **66**; 255–266.

Gieryng R. 1965. Veränderungen der histologischen Struktur des Gehirns von *Calliphora vomitoria* (L.) (*Diptera*) während der postembryonalen Entwicklung. *Z. wiss. Zool.*, **171**; 80–96.

Girardie J., Girardie A. 1972. Evolution de la radioactivité des cellules neurosécrétrices de la *pars intercerebralis* chez *Locusta migratoria migratorioides* (*Insecta Orthoptera*) après injection de cystéine S^{35}. Étude autoradiographique aux microscopes optique et électronique. *Z. Zellforsch. mikrosk. Anat.*, **128**; 212–226.

Gorbman A., Bern H. A. 1966. Neurosecretion and neuroendocrine relations in the invertebrates. [In:] *A textbook of Comparative Endocrinology*, pp. 377–404. John Wiley and Sons, Inc. New York, London, Sydney.

Grillot J. P. 1977. Les organes périsympathiques des diptères. *Int. J. Insect Morphol. Embryol.*, **6**; 303–343.

Harker J. E. 1960. Endocrine and nervous factors in insect circadian rhythms. *Cold Spring Harb. Symp. Quant. Biol.* **25**; 279–287.

Herman W. S., Gilbert L. J. 1965. Multiplicity of neurosecretory cell types and groups in the brain of the saturniid moth *Hyalophora cecropia* (L.). *Nature*, **205**; 926–927.

Herman W. S., Gilbert L. J. 1966. The neuroendocrine system of *Hyalophora cecropia* (L.) (*Lepidoptera*: *Saturniidae*). I. The anatomy and histology of the ecdysial glands. *Gen. Comp. Endocrinol.*, **7**; 275–291.

Hertel W., Koch J., Penzlin H. 1978. Elektrophysiologische Untersuchungen am Frontalganglion von *Periplaneta americana* L. *J. Insect Physiol.*, **24**; 721–735.

Highnam K. C., Mordue A. J. 1970. Estimates of neurosecretory activity by an autoradiographic method in adult female *Schistocerca gregaria* (Forsk.). *Gen. Comp. Endocrinol.*, **15**; 31–38.

Isobe M., Neswegawa K., Kubota J., Gotto T. 1976. Diapause hormone B: Its selective extraction and isolation from the silkworm, *Bombyx mori*. *Agr. Biol. Biochem.*, **40**; 1189–1199.

Jawłowski H. 1960. On the brain structure of the *Symphyta* (*Hymenoptera*). *Bull. Acad. Pol. Sci.*, **8**; 265–268.

Johansson A. S. 1958. Neurosecretion in the milkweed bug, *Oncopeltus fasciatus*. II. *Intern. Symp. Neurosekr. Lund.*, pp. 98–101. Springer-Verlag, Berlin.

Kind T. W. 1978. Neurosecretory cells of the brain of *Mamestra brassicae* (*Lepidoptera*, *Noctuidae*) during metamorphosis and diapause development. [In:] *Neurosecretion and Neuroendocrine Activity*, pp. 378–381. Springer-Verlag, Berlin, Heidelberg, New York.

King R. C., Aggarwal S. K., Bodenstein D. 1966. The comparative submicroscopic morphology of the ring gland of *Drosophila melanogaster* during the second and third larval instars. *Z. Zellforsch. mikrosk. Anat.*, **73**; 272–285.

Lee H. T.-Y. 1948. A comparative morphological study of the prothoracic glands of some lepidopterous larvae, with special reference to their innervation. *Ann. entomol. Soc. Amer.*, **41**; 200–205.

Lüscher M. 1960. Hormonal control of caste differentiation in termites. *Ann. Acad. Sci. N.Y.* **89**; 549–563.

Mahon D. C., Nair K. K. 1975. A comparison of aldehyde fuchsin and alcian blue staining of neurosecretory material in *Oncopeltus fasciatus. Cell. Tiss. Res.*, **161**; 477–484.

Malá J., Novák V. Y. A., Blazsek J., Balázs A. 1974. The effects of juvenile hormone on the prothoracic glands in *Galleria mellonella* L. I. Morphology of the glands in the course of the postembryonic development. *Acta biol. Acad. Sci. hung.*, **25**; 85–95.

Marks E. P., Holman G. M., Borg T. K. 1973. Synthesis and storage of a neurohormone in insect brains *in vitro. J. Insect Physiol.*, **19**; 471–477.

Morohoshi S., Mori T., Sato S. 1976. The control of growth and development in *Bombyx mori*, XXXIII. Electron-microscopic studies of the activity of *corpus allatum* in the fifth instar larvae developed from low and high temperature eggs. *Proc. Japan Acad.*, **52**; 438–441.

Muszyńska-Pytel M., Cymborowski B. 1978. The role of serotonin in regulation of the circadian rhythms of locomotor activity in the cricket (*Acheta domesticus* L.). II. Distribution of serotonin and variations in different brain structures. *Comp. Biochem. Physiol.*, **59** C; 17–20.

Nijhout H. F. 1975. Axonal pathways in the brain-retrocerebral neuroendocrine complex of *Manduca sexta* (L.) (*Lepidoptera*: *Sphingidae*). *Ins. J. Insect Morphol., Embryol.*, **4**; 529–538.

Novák V. J. A. 1975. *Insect Hormones*, pp. 82 and 121. Chapman and Hall, London.

Odhiambo T. R. 1966. The fine structure of the *corpus allatum* of the sexually mature male of the desert locust. *J. Insect Physiol.*, **12**; 819–828.

Pipa R. L., Novák F. J. 1979. Pathways and fine structure of neurons forming the *nervi corporis allati II* of the cockroach *Periplaneta americana* (L.). Cell Tissue Res., **201**; 227–237.

Provansal A., Grillot J. P. 1972. Les organes périsympathiques des insectes holométaboles. I. Coléoptères. *Ann. Soc. entomol. Fr.* (*N.S.*), **8**; 863–913.

Przełęcka A., Cymborowski B. 1970. Znaczenie funkcjonalne komórek neurosekcyjnych u owadów [Functional significance of neurosectory cells in insects]. *Kosmos A*, **1**; 43–58.

Raabe M. 1965. Étude des phénomènes de neurosécrétion an niveau de la chaine nerveuse ventrale des Phasmides. *Bull. Soc. Zool. France*, **90**; 631–654.

Raabe M. 1967. Recherches récentes sur la neurosécrétion dans la chaine nerveuse ventrale des insectes. *Bull. Soc. Zool. France*, **92**; 67–71.

Raabe M., Monjo D. 1970. Recherches histologiques et histochimiques sur la neurosécrétion chez le Phasme, *Clitumnus extradentatus*: les neurosécrétions de type. *C.R. hebd. Séanc. Acad. Sci., Paris*, **270**; 2021–2024.

Raabe M., Baudry N., Grillot J. P., Provansal A. 1974. Neurosecretion—The Final Neuroendocrine Pathway. VI. *International Symposium on Neurosecretion. London* **1973**, pp. 59–71. Springer-Verlag, Berlin, Heidelberg, New York.

Rowell H. F. 1976. The cells of the insect neurosecretory system: Constancy, variability, and the concept of the unique identifiable neuron. *Adv. Insect Physiol.*, **12**; 63–123.

Scharrer B. 1945. Neurosecretion. XI. The effects of nerve section on the *intercerebralis-cardiacum-allatum* system of the insect *Leucophaea maderae. Physiol. Rev.*, **25**; 261–272.

Scharrer E., Scharrer B. 1954. Hormones produced by neurosecretory cells. *Recent Progress Hormone Res.*, **10**; 183–232.

Scharrer B. 1963. Neurosecretion. XIII. The ultrastructure of the *corpus cardiacum* of the insect *Leucophaea maderae*. *Z. Zellforsch. mikrosk. Anat.*, **60**; 761–796.

Scharrer B. 1971. Histophysiological studies on the *corpus allatum* of *Leucophaea maderae*. V. Ultrastructure of sites of origin and release of a distinctive cellular product. *Z. Zellforsch. mikrosk. Anat.*, **120**; 1–16.

Schooneveld H. 1970. Structural aspects of neurosecretory and *corpus allatum* activity in the adult colorado beetle, *Leptinotarsa decemlineata* Say, as a function of daylength. *Netherlands J. Zool.*, **20**; 151–237.

Schooneveld H., Sanchez O., de Wilde J. 1977. Juvenile hormone-induced break and termination of diapause in the colorado potato beetle. *J. Insect Physiol.*, **23**; 689–696.

Schreiner B. 1966. Histochemistry of the A cell neurosecretory material in the milkweed bug, *Oncopeltus fasciatus* Dallas (*Heteroptera*: *Lygaeidae*), with a discussion of the neurosecretory material carrier substance problem. *Gen. Comp. Endocrinol.*, **6**; 388–400.

Singh H. H., Sehnal F. 1979. Lack of specific neurons in the ventral nerve cord for the control of prothoracic glands. *Experientia*, **35**; 1117–1119.

Sláma K., Romaňuk M., Šorm F. 1974. *Insect Hormones and Bioanalogues*. Springer-Verlag, Wien, New York.

Smith U., Smith D. S. 1966. Observations on the secretory processes in the *corpus cardiacum* of the stick insect, *Carausius morosus*. *J. Cell. Sci.*, **1**; 59–66.

Thomsen E., Thomsen M. 1970. Fine structure of the *corpus allatum* of the female blow-fly *Calliphora erythrocephala*. *Z. Zellforsch. mikrosk. Anat.*, **110**; 40–60.

Tombes A. S., Smith D. S. 1970. Ultrastructural studies on the *corpora cardiaca-allata* complex of the adult alfalfa weevil, *Hypera postica*. *J. Morphol.*, **132**; 137–148.

Waku Y., Gilbert L. J. 1964. The *corpora allata* of the silkmoth, *Hyalophora cecropia*: An ultrastructural study. *J. Morph.*, **115**; 69–96.

Wartoń S., Dutkowski A. B. 1978. Ultrastructural analysis of the action of reserpine on the brain neuroendocrine system of the wax moth, *Galleria mellonella* L., *Lepidoptera*. *Cell. Tiss. Res.*, **192**; 143–155.

Weber W., Gaude H. 1971. Ultrastruktur des Neurohaemalorgans im *Nervus corporis allati II* von *Acheta domesticus*. *Z. Zellforsch. mikrosk. Anat.*, **121**; 561–572.

Weyer F. 1935. Ueber drüsenartige Nervenzellen im Gehirn der Honigbiene, *Apis mellifica* L. *Zool. Anz.*, **112**; 137–141.

Wigglesworth V. B. 1935. Function of the *corpus allatum* in insects. *Nature*, **136**: 338.

Wilde J., de 1968. Hormones and diapause. [In:] *Progress in Endocrinology*, pp. 351–355, Excerpta Medica International Congress, Vol. 184.

Wilde J., de, de Boer J. A. 1969. Humoral and nervous pathways in photoperiodic induction of diapause in *Leptinotarsa decemlineata*. *J. insect Physiol.*, **15**; 661–675.

Williams C. M. 1969. Nervous and hormonal communication in insect development. *Developmental Biol. Suppl.*, **3**; 133–150.

Chapter 3

Methods and techniques in insect endocrinology

The methods used to study the insect endocrine system are similar to those used in the endocrinology of vertebrates. This is quite natural since, as demonstrated by Scharrer and Scharrer (1944), there are close analogies in the structure and function of the basic endocrine centres in these two distant groups of animals.

In general these methods consist of: (1) identification of the individual centres of the endocrine system, and (2) investigation of their function. In attempts to determine whether a given organ or cell group has endocrine functions, the usual method is to eliminate and re-implant the organ or cell group. The centre in question is normally removed surgically, and the consequences of this are studied. If reimplantation of the tissue or organ that was removed restores normal function, we may be sure that the centre participates in endocrine control.

A much more complicated method is required when studying the function of a particular endocrine organ. This problem is complicated by the fact that the insect endocrine system constitutes a uniform functional whole, and it is very difficult to investigate the activity of any one of its parts.

In spite of these difficulties, there are at present numerous methods available for the study of the structure and function of the insect endocrine system. Some of them will be described in this chapter.

3.1. Elimination of insect endocrine centres

Removal of an endocrine organ deprives the organism of its natural source of a hormone or hormones. As a consequence of this, disturbances are observed in physiological processes. The oldest method of eliminating insect endocrine glands is to fit ligatures which prevent the passage of hormones into the haemolymph. As already mentioned, this method was used for the first time by Kopeć in 1917 and is still in common use (Sláma *et al.*, 1974, pp. 124–129). Since the chief endocrine glands of insects—both larvae and imagines—are in the head segment, it is possible to isolate them from the remaining part of the body by placing a ligature between the head capsule and the thorax (Plate 8A). The brain, suboesophageal ganglion, *corpora allata*, *corpora cardiaca* and frontal ganglion are thus separated from other parts of the body. The endocrine centres in the thorax (prothoracic glands) can be isolated by placing a ligature between the thorax and the abdomen (Plate 8 B). Only hormones produced by neurosecretory cells located in the abdominal ganglia will then be released into the haemolymph.

Although the method described above has contributed to the elucidation of many endocrinological problems in insects, it has serious drawbacks. In the first place the ligature deprives the insect of the ability to feed, and this, as we know, may have a decisive influence on the functioning of the links within the endocrine system. Ligatures should therefore be fitted, if at all possible, when the larvae are entering the phase of so-called physiological starvation in the period preceding pupation. Frequently, however, the experiment requires that the endocrine glands be eliminated during a period of feeding. In such cases the glands must be removed. Owing to the convenient location of most endocrine centres, this is possible. For instance, the entire brain can be removed leaving the suboesophageal ganglion with its neurosecretory cells intact (Cymborowski, 1970a), or the neurosecretory cells of the brain can be selectively destroyed by thermocauterization (Cymborowski, 1970a) or electromagnetic radiation of a specific wavelength (Cymborowski, 1973b).

3.1.1. Removal of neurosecretory cells

In investigations of the function of the insect endocrine system it is frequently necessary to remove only the neurosecretory cells. Since these cells exhibit a characteristic opaque fluorescence (see Chapter 2), they are easy to observe *in situ* under the stereoscopic microscope after exposure of the brain (Plate 9), and they can also be removed mechanically. For this purpose the insect should be anaesthetized.

The most commonly used way of anaesthetizing insects is to expose them for a few minutes to an atmosphere of carbon dioxide (Williams, 1946b). Numerous experiments have shown, however, that if the CO_2 concentration reaches 70 per cent of the gas mixture it may inhibit growth and development (Brooks, 1957; Edwards and Patton, 1965). The range of safety is from 20 to 30 per cent of CO_2, and to maintain the concentration within these limits it is necessary to control the gas flow using a special device, a simplified version of which has been described by Patton *et al.* (1968).

To avoid the inconveniences associated with the use of CO_2 for anaesthesia, increasing use has recently been made of so-called water anaesthesia. Immersion of the insect in water for 10–15 min caused reflex closure of the spiracles, so that the animal is temporarily poisoned by metabolic CO_2. When removed from the water and dried on filter paper the insect returns to a normal state after a few minutes. This type of anaesthesia eliminates the danger of CO_2 overdosage, and, more importantly, does not affect further development and metamorphosis.

In experiments with the cricket (*Acheta domesticus*) involving the removal of neurosecretory cells from the brain (Cymborowski, 1970a), or the injection of various substances directly into the brain (Cymborowski, 1972a) a specially adapted container for the immobilization of the insect is placed on a revolving table (Fig. 3.1). There is a slot in the lid of the container into which is fitted the

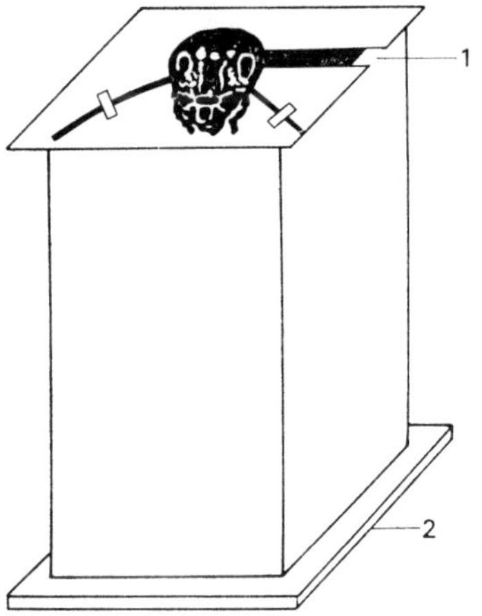

Figure 3.1. Preparation of insect for removal of brain neurosecretory cells.
1—slot in lid of container, 2—revolving table facilitates placement of insect's head under the dissection microscope. The container is filled with water in which the insect is immersed, keeping it under continuous anaesthesia.

neck segment of the insect, so that its head can be conveniently attached with Plasticine to the upper surface of the lid. The remaining part of the insect's body is immersed in water, allowing anaesthesia during the course of microsurgery.

Using this preparation, the insect's brain can be exposed (Plate 10). The first step, after sterilization of the head with 70 per cent ethanol, is to make an opening (1.5 × 1.5 mm) in the head with a very thin blade. The cuticle removed should be put aside on a sterilized glass plate and used to cover the hole after removal of the neurosecretory cells.

The perineural lamina of the brain is then punctured at the site where the neurosecretory cells are located (see Plate 9). The cells will flow onto the brain surface, and it will be relatively easy to remove them using a bent dissection needle. This method was successfully used in the experiments on crickets referred to above (Cymborowski, 1970a).

Neurosecretory cells in the brain can also be destroyed with an electric current (Cymborowski, 1970). For this, however, a special metal or glass electrode is necessary. An appropriate electrode is also indispensable for the cauterization of neurosecretory cells with electromagnetic waves (Cymborowski, 1973a). In these experiments a very simple ultrashort-wave generator was utilized.

After destruction of the neurosecretory cells, the opening in the head capsule should be carefully closed. For this purpose paraffin or wax with low melting point are usually used.

All the methods described require control experiments. In the first place their effectiveness has to be ascertained, and the site and area of the part of the brain that has been destroyed must be determined. When cell destruction is only partial, or when brain centres other than those targeted are damaged, the preparation should be discarded. The accuracy of cell destruction can be ascertained after the end of the experiment, by examination of stained histological preparations of the brain.

A control experiment consisting of a sham operation is performed to determine the effect of the lesion itself on the insect's behaviour and on the vital processes studied. Thus, when cauterizing the neurosecretory cells, a second group of insects should be subjected to all the steps of the operation except cell destruction. The two groups of insects should then be compared to determine the degree to which the operation itself disturbs the animal's normal functions.

Cauterization of neurosecretory cells is rather difficult to perform and the results obtained are frequently not unequivocal (Cymborowski, 1973a). This is because it is very difficult to ascertain whether all cells or only some of them have been removed by surgery. Cells remaining after cauterization may take over the function of the destroyed cells, which greatly complicates the picture of changes obtained. Moreover, the cells of the insect nervous system may regenerate (Jacklet and Cohen, 1967; Cymborowski, 1968), which leads to discrepancies in the results depending on the time that has elapsed since the operation (Fig. 3.2).

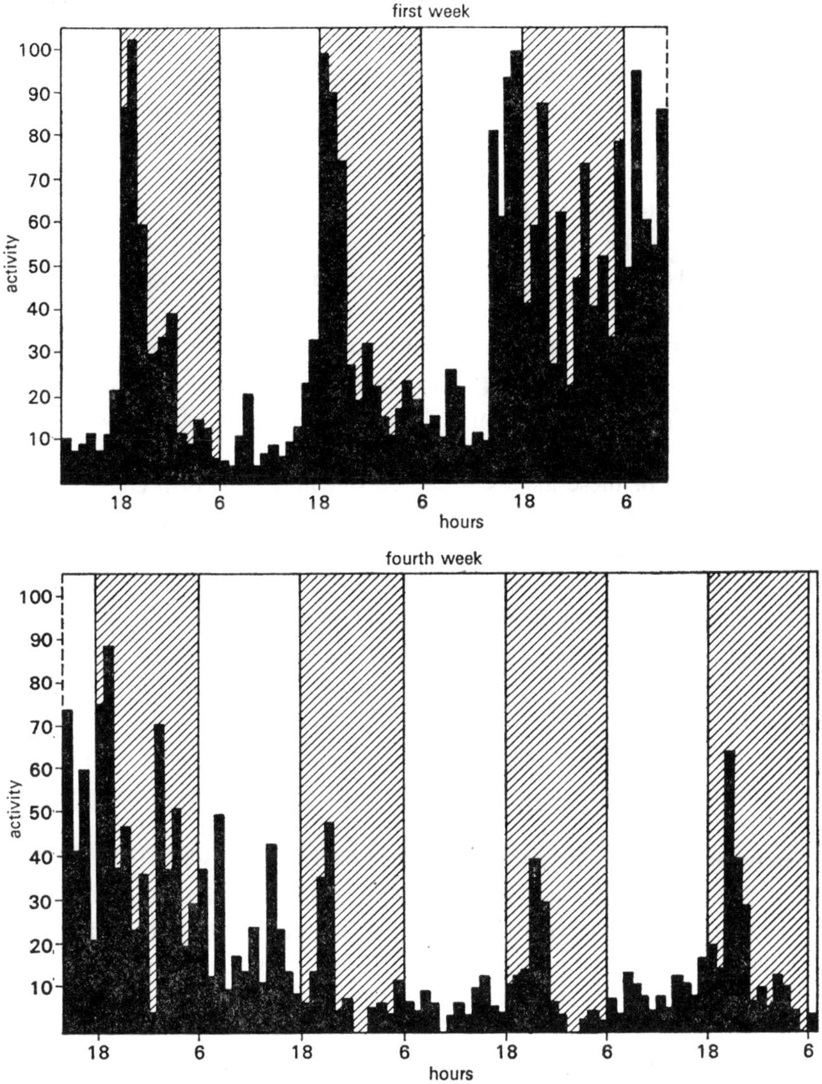

Figure 3.2. Effect of removal of neurosecretory cells of *pars intercerebralis* from the brain of *Acheta domesticus* on the rhythm of motor activity. Note that the rhythm is restored in the fourth week after operation. This example indicates that the neurosecretory cells may regenerate. Black bars indicate the number of movements per hour, hatched areas—periods of darkness.

3.1.2. Removal of insect endocrine glands

Many now classical experiments in insect endocrinology have been performed on the reduviid bug *Rhodnius prolixus* (*Hemiptera*) (Wigglesworth, 1952a, b; 1955). It is a parasitic species feeding on the blood of man and other mammals. The elongated head of this insect is a characteristic feature, and the shape of the head makes it possible to selectively remove the neurosecretory cells of the protocerebrum after partial decapitation of the animal. When the incision is made somewhat nearer to the thorax, the *corpora allata* (*CA*), the source of juvenile hormone, can also be removed. This is a very simple way of removing these endocrine glands, though it can be used only in *Rhodnius prolixus*. The shortcoming of this technique is the necessity of removing the neurosecretory cells whenever the *CA* are removed, making its impossible to study the function of the *CA* alone. The technique of removing the *corpora allata* and *corpora cardiaca* (*CC*) developed for *Pyrrhocoris apterus* by Sláma (1964) is therefore worth mentioning.

The removal of these glands is carried out in a Petri dish containing insect Ringer solution with small amounts of penicillin and streptomycin added. To expose the *corpora allata* and *corpora cardiaca* a transverse section is made through the neck of the insect. Then, using a thin, hooked dissection needle the glands are separated from each other, and the *CA* or *CC* are removed with a microforceps (each type of gland separately). The mortality of insects after this treatment is low, usually not exceeding 10 per cent. Immersion in physiological saline is at the same time a way of anaesthetizing the insect, and facilitates work. It should be stressed here that after the end of each experiment autopsy must be performed to confirm the accuracy of the surgery. Control experiments are also necessary.

Many simple methods have been developed for removing the *CA* and *CC*. Among others Sehnal and Granger (1975) give a description of *CA* removal in wax moth (*Galleria mellonella*) larvae. In studies of the function of the *CA* during metamorphosis of *Manduca sexta* larvae these organs were removed using new methods (Cymborowski, unpublished data; Kiguchi and Riddiford, 1978).

Although the technique of *CA* and *CC* removal is relatively simple and can be applied to most insect species (e.g. Strong, 1963, designed a simple device for *CA* removal in the locust), removal of the prothoracic glands is practically impossible. This is due to the inconvenient position of these glands as well as their structure. In most insect groups they consist of a large number of single cells, frequently loosely dispersed over a wide area in the prothorax (see Chapter 2).

The removal of endocrine glands using the methods described, allows the study of many aspects of insect endocrinology, since it offers the possibility of effecting a gradual reduction in hormone titre, for instance by removal of the *CA* or *CC*.

3.2. Implantation of endocrine glands

To demonstrate that a given organ has endocrine functions it is necessary to implant it into the body of a recipient from which the organ has previously been removed. The restoration of normal physiological processes after reimplantation means that the organ in question influences the animal's physiology via substances released into the haemolymph. Such experiments are very frequent in endocrinology.

Doane (1961) demonstrated that it is possible to transplant endocrine glands (*corpora allata* and *corpora cardiaca*) between various *Drosophila melanogaster* mutants, since histological examination of the implants indicated survival and normal function of the cells for up to one week after transplantation.

In experiments on the role of the neurosecretory cells of the *pars intercerebralis* of the cricket brain in the control of the diurnal rhythm of locomotor activity, brains from rhythmic donors were transplanted into the body cavity of recipients with nonrhythmic activity. The nonrhythmic insects were obtained by cauterizing the neurosecretory cells of the *pars intercerebralis*, then implanting brains with intact neurosecretory cells into the abdomen. It appeared that brains from rhythmic donors periodically secrete a substance inhibiting locomotor activity into the haemolymph of nonrhythmic recipients (Cymborowski, 1968).

Instead of implantation of the whole brain, it is possible to implant a part of it such as the *pars intercerebralis* or the optic lobes, or even secretory cells prepared as described on page 44.

Active glands can also be implanted as a means of raising the titre of a particular hormone in the haemolymph in order to observe the effects of the hormone. An example of this in common use is the implantation into the body cavity of a complex of several active centres, such as the brain, *CC* and *CA*, using the method of Piepho (1940).

This method was used for the first time in experiments with wax moth (*Galleria mellonella*) larvae. It consists of taking endocrine glands from one or several larvae and introducing them into the body cavity of another larva. Implantation is done under anaesthesia and the tissues are introduced through an incision made in the abdominal segment. In this way the recipient has additional, efficiently functioning endocrine centres as well as its own. The same technique may be used in the separate implantation of *corpora allata* and *corpora cardiaca*. It is also possible to implant prothoracic glands, parts of these glands, or particular ganglia.

As in experiments involving the cauterization of neurosecretory cells, control experiments should be carried out in which the insect is merely wounded at the site of implantation.

3.3. Parabiosis

The simplest way of transfusing haemolymph from one insect to another is by means of parabiosis. This consists of connecting the body cavities of two or more insects with thin tubes, so as to allow a free flow of heamolymph between the insects. The well-known experiment of Wigglesworth (1958) used this technique to investigate the hormonal control of growth and metamorphosis in *Rhodnius prolixus*. Earlier the same author, by connecting decapitated larvae in various developmental instars, demonstrated that the decisive event in the development of these insects is the intake of a critical amount of food—mammalian blood (Wigglesworth, 1957).

Parabiosis has also been used to study the hormonal regulation of the diurnal rhythm of locomotor activity in insects (Cymborowski and Brady, 1972). In this study two insects (crickets or cockroaches) were connected one above the other; the lower one was able to move, but all its retrocerebral glands were eliminated by decapitation; the insect lying above had only its legs removed. The two insects had a common circulatory system, obtained by passing a thin polyethylene tube through the site where their prothoraxes joined (Plate 11). The lower insect was arrhythmic as regards locomotor activity, whereas the upper one was rhythmic. To eliminate other causes of rhythmicity in the nonrhythmic insect, for instance periodic movements executed by the rhythmic insect, control parabiosis was obtained by the use of an occluded tube, so that neither insect's haemolymph could pass to the other insect. The results of these experiments will be discussed in Chapter 8.

3.4 Culture of endocrine glands *in vitro*

Since the first experiments of Richard Goldschmidt on insect tissue culture half a century ago, the technique has been used intensively in the study of insect morphogenesis (Day and Grace, 1959). However, the method was not applied immediately in physiological research. It was considered that the culture of cells and tissues outside the insect body would lead to drastic changes which would preclude study of their normal functions. It seems more likely, however, that the lack of interest in tissue culture is related to the great diversity of insect species, and the consequent necessity for tedious and time-consuming investigations of many different kinds of haemolymph in order to obtain physiological solutions of appropriate composition, pH and osmotic pressure. There is available at present a whole spectrum of physiological solutions, developed for many insect groups. The composition of some of them is listed in the papers of Martignoni (1962).

With the physiological salines now available, a wide range of *in vitro* studies on the function of the insect endocrine system is possible (Marks, 1971, Seshan, 1976). This technique makes it possible to study the metabolism of the isolated organ; the functioning tissue secretes into the environment metabolites of various types, the nature of which will indicate the current physiological state of the tissue (Marks and Holman, 1974). The greatest advantage of this method, however, is that it makes possible investigation of the interaction between individual sources of hormones in the insect endocrine system. For instance, the *in vitro* incubation of various tissues of *Acheta domesticus* (Dutkowski and Cymborowski, 1973) established that the suboesophageal ganglion, by releasing into the incubation medium a chemical factor of unknown nature, inhibits metabolic processes occurring within the fat body cells. Thus, the *in vitro* incubation of endocrine organs makes it possible to ascertain the chain of dependencies between particular organs and tissues.

Use of the *in vitro* culture technique combined with other methods permits, for instance, investigations on exygen uptake by the brain, *CA*, *CC* or prothoracic glands (Cymborowski, 1975). Thus it is possible to investigate the functions of endocrine sources at various instars of insect development.

By adding insect hormones to the incubation fluid we are able to study the mechanisms of action of the endocrine system. Marks *et al.* (1972), for instance, by adding ecdysone at a concentration of 2.5 μl/ml to medium containing brains of *Leucophaea maderae* larvae, found that this hormone causes release of neurosecretion from the neurosecretory cells of the *pars intercerebralis*. The results of this experiment suggest, therefore, the existence of a feedback between the neurosecretory cells of the insect brain and the prothoracic glands, producing ecdysone.

Tissue culture has now progressed so far that the long-term culture of single cells of the insect neurosecretory system (Plate 12) or of whole endocrine glands is possible. The only problem is that prolonged incubation of endocrine glands causes such severe changes in their morphology that they cease to resemble the glands that function normally in the insect body. For instance, the *corpora cardiaca*, which have a definite structure and location in the body, produce innumerable nerve processes after 40 days of *in vitro* culture (Plate 13). The physiology of the organ probably also undergoes great changes during prolonged incubation.

3.5 Histochemical methods

Histochemical methods are extremely valuable in investigations of the function of the insect endocrine system. These techniques are utilized in making histological preparations, and for appropriate staining of the cells constituting the endocrine tissues.

3.5.1. Methods of staining neurosecretory cells

One of the most commonly used ways of staining the neurosecretory cells of the insect endocrine system is the Gomori method (1950), in which microtome sections are stained with aldehyde fuchsin. This method was originally used to stain connective tissue, but it was later adapted for neurosecretory cells of vertebrates and invertebrates by Halmi (1952) and Cameron and Steele (1959).

Extremely important, and decisive for the success of this method, is suitable preparation of the dye—aldehyde fuchsin. It should be prepared as follows: dissolve 1 g alkaline fuchsin in 200 ml of boiling distilled water, and boil for about 1 min. Then cool the solution and pass it through filter paper. Carefully add 2 ml of concentrated HCl and 2 ml paraldehyde to the filtrate. The solution thus prepared should be left to stand at room temperature for 3–4 days to mature. With time, the amount of fuchsin in the solution will gradually diminish and the amount of sediment increase. This may be checked by placing a drop of the solution on filter paper. After 3–4 days the solution should be filtered and the filtrate discarded. The sediment remaining on the filter paper should be dried at about 50°C and stored in a closed vessel. In this way about 2 g of dry dye is obtained.

To stain neurosecretory cells the solution should be prepared as follows: dissolve 250 mg of the aldehyde fuchsin powder in 50 ml of 70 per cent ethanol. The stain thus prepared remains usable for about 6 months.

The tissue is fixed in Bouin's solution for at least 24 h and after dehydration it is embedded in paraffin. Then 5-μm microtome sections are prepared.

The process of staining is as follows:

1. Deparaffinize the sections by passing through xylene and absolute ethanol + xylene 1:1, then rehydrate by immersing successively in a decreasing series of ethanol concentrations (100%, 95%, 80%, 70%, 50%, 30%, H_2O).

2. Oxidize for 1 min in Gomori solution, prepared by dissolving 150 mg $KMnO_4$ in 50 ml distilled water acidified with 0.1 ml concentrated sulphuric acid. This solution should be freshly prepared before use.

3. Immerse sections for several seconds in a 2.5 per cent $NaHSO_3$ solution, until colour disappears.

4. Wash with distilled water.

5. Transfer sections to 70 per cent ethanol via intermediate 30 and 50 per cent solutions.

6. Stain in aldehyde fuchsin solution for 2–10 min.

7. Dry the sections rapidly with filter paper and transfer to 95 per cent ethanol.

8. Transfer sections to another vessel containing 95 per cent ethanol and leave for 2–5 min, until alcohol ceases to take up colour.

9. Hydrate sections after immersion in 70 and 30 per cent ethanol.

10. Stain for 20–45 s in Halmi's mixture. Prepare it by adding successively 100 ml distilled H_2O, 0.2 g light green, 1 g orange G, 0.5 ml phosphotungstic acid and 1 ml glacial acetic acid.

11. Dry sections with filter paper and counterstain them for no longer than 1–2 min in 95 per cent ethanol acidified with several drops of 0.2 per cent acetic acid.

12. Transfer sections to 95 per cent ethanol.

13. Wash twice with 100 per cent ethanol, clarify with xylene and seal in balsam or DePeX.

This method is very sensitive, allowing the detection of differences in the accumulation of neurosecretory material in cells (Plate 14). Depending, on the amount of material accumulated, either the cytoplasm of the cells stains an intense dark blue (Plate 14 A) or the neurosecretion appears in the form of loosely dispersed granules in the cytoplasm (Plate 14 B). According to some authors, however, aldehyde fuchsin stains the transporting proteins of the neurosecretory cells; the biologically active components of the neurosecretion can be stained with alcian blue 8 GX (Mahon and Nair, 1975). This stain is one of the alkaline copper phthalocyanates, and its use to stain neurosecretory cells is described by Schoumacker and van Damme (1971).

Results similar to those given by aldehyde fuchsin are also obtained with aldehyde thionine, using Paget's method (1959). The stain is prepared by mixing the following reagents: 0.5 g thionine, 91.5 ml 70 per cent ethanol, 7.5 ml paraldehyde and 1 ml concentrated hydrochloric acid. Oxidation of tissues is done in the same way as with aldehyde fuchsin staining. Bleaching of the sections, however, is carried out in 2 per cent $NaHSO_3$ for 2–10 min. The duration of staining must not be less than 15 min, and should not exceed 2 h.

Differentiation is in distilled water, and after dehydration the samples are embedded in balsam or DePeX.

It is also possible to stain the neurosecretory cells selectively in whole-tissue preparations of the insect endocrine system by staining *in situ* using Dogra and Tandan's method (1964). In this method, the brain and endocrine glands of the anaesthetized insect are first exposed, and the tissues are then fixed in 10 per cent formaldehyde in insect saline. The fixed tissues are then removed and stained as a whole with Victoria blue.

This method reveals the spatial distribution of the neurosecretory cells in the tissue, and makes it possible to follow the course of the axons through which the neurosecretion travels.

3.5.2. Cobalt technique

Another method for visualizing the spatial distribution of neurosecretory cells and the innervation of endocrine glands is the cobalt technique. This method was developed and improved comparatively recently (Hang, 1973) and is now widely used in insect endocrinology. Modifications of the method are presented in Tyrer and Bell's paper (1974, p. 151). It consists of the *in vivo* introduction into the nervous system of cobalt chloride ($CoCl_2$), which is then precipitated in the form of black cobalt sulphide under the action of ammonium sulphide. This method allows even the smallest branches of the nerve cells to be visualized (Truman and Reiss, 1976). The greatest advantage of the method is the possibility of observing the course of neurons throughout the entire organ and also in histological sections after fixation of the tissue in Bouin's solution. A method has also been devised for the preparation of slides for electron microscopy. For this purpose staining of the nerve cells is intensified by the use of Timm developer containing silver nitrate (Pihl and Falkner, 1967). A disadvantage of this method is the necessity of keeping the issue alive to increase its saturation with cobalt salts.

3.5.3. Autoradiographic methods

Autoradiographic methods allow the current physiological condition of the endocrine centres in the insect to be studied, and are used increasingly in studies of the mechanisms of action of insect hormones. The theoretical basis of autoradiography and various methods used in this field are described in the excellent monograph by Rogers (1973) and in a paper by Sawicki (1963).

Autoradiographic methods are particularly useful in studies of insect neurosecretory cells (Cymborowski, 1968, 1970b; Cymborowski and Dutkowski, 1969). A good index of the physiological condition of these cells is the level of RNA synthesis. Nucleic acid synthesis can be investigated by autoradiography using tritium-labelled uridine as precursor. This can be performed *in vivo* by injecting ^{3}H-uridine dissolved in physiological saline into the haemolymph, or *in vitro* by adding the labelled precursor to the medium in which the tissue is placed.

It is extremely important to adjust the concentration of the isotope so as not to disturb cell metabolism, and at the same time allow the isotope to be detected by the autoradiographic emulsion, or by Pelc's method of scraping the film (Pearse, 1960). In this type of study concentrations of the order of 10 μCi/g body weight are used.

The quality of the autoradiograms also depends on the choice of a suitable incubation period, i.e. the interval from administration of the labelled substance to fixation of the tissue. The distribution of RNA that has incorporated ^{3}H-uridine within the cell is also dependent on this period. In autoradiographic investigations of RNA synthesis in brain neurosecretory cells and suboesophageal ganglion cells of the cricket (Cymborowski, 1970b) three stages were noted in distribution of labelled RNA (Plate 15).

An autoradiographic study of the rate of RNA synthesis was also performed in cells of the prothoracic glands (Oberlander *et al.*, 1965). Uridine at concentrations of 5 or 10 μCi/g body weight was used as a precursor of RNA. The insects were killed 1, 2 or 12 h after injection of the radioactive substance.

To ensure that it is RNA synthesis that is being estimated on the basis of the autoradiograms, a series of controls should be run in which the preparations are digested with ribonuclease before application of the emulsion. If, after such treatment, darkening of the autoradiographic emulsion (characteristic of RNA) does not occur, the ^{3}H-uridine has in fact been incorporated into the RNA.

Another autoradiographic technique for studying the physiological condition of neurosecretory cells involves the use of a mixture of labelled amino acids

(Cymborowski and Dutkowski, 1970) as protein precursors. Since there is a high percentage of cysteine in the neurosecretory material, best results are obtained using ^{35}S-cysteine as precursor (Cymborowski, 1973b).

The use of labelled cysteine not only allows the intensity of synthesis in the neurosecretory cells to be determined, but also enables the translocation of neurosecretory material along the axons and its accumulation in the *corpora cardiaca* or *corpora allata* to be observed (Plate 16). It is also possible to run the experiments so as to determine the rate of release of neurosecretory material into the haemolymph (Highnam, 1961).

3.6. Pharmacological methods

Many chemical substances, especially alkaloids, have proved useful in studies of the insect endocrine system. One of them is reserpine, which when injected into the insect haemolymph inhibits the secretory functions of the brain's neurosecretory cells (Cymborowski, 1970; Masner *et al.*, 1970). The ability to arrest the physiological activity of these cells makes it possible to study the functioning of the insect neurosecretory system.

In investigations on the cricket (Cymborowski, 1973a) and the wax moth (Cymborowski and Kryspin, 1975) it has been demonstrated that injection of reserpine into the body cavity blocks the release of neurosecretory material from the neurosecretory cells of the *pars intercerebralis*. Several days after the administration of reserpine, the amount of neurosecretory material that had accumulated in the neurosecretory cells was much greater than that observed in control insects treated with physiological saline. At the same time, it was found, by the use of autoradiographic methods, that RNA synthesis in the neurosecretory cells of the *pars intercerebralis* was greatly reduced after injection of reserpine. Thus, we have available a relatively selective method of blocking the function of neurosecretory cells, which may in many cases make it unnecessary to remove these cells surgically.

3.7. Methods of assaying hormone activity

On account of the extremely low concentrations at which hormones occur in insects, chemical assay methods are not generally used. A number of highly sensitive bioassays have therefore been developed, and these are frequently used in insect endocrinology. This section will describe some of the commonly used bioassays, including details of methods of extracting the hormones and determining their biological activity.

3.7.1. Evaluation of prothoracicotropic hormone (PTTH) activity

The prothoracicotropic hormone (PTTH) is still a little-known protein. Because of this it is difficult to devise suitable methods of determining its activity in the insect body. Methods for the partial purification of the hormone were devised relatively recently (Suzuki *et al.*, 1975; Yamazaki and Kobayashi, 1969). This was an indispensable condition for progress in investigations on this hormone. The purification procedure of PTTH from the *Bombyx mori* brain is as follows:

Silkworm heads
(more than 100 000 heads)
↓
acetone powder
↓
washing with 80% ethanol
↓
extraction with 2% aqueous NaCl solution (3 times)
↓
precipitation with 80% $(NH_4)SO_4$
↓
precipitation with 50–75% acetone
↓
precipitation with 90% picric acid
↓
precipitation with 80% acetone
↓
crude PTTH extract.

Further purification is more complicated, but yields PTTH of high purity. Purification is as follows:

crude PTTH extract
↓
Sephadex G-50
(3 times)
↓
DEAE-sepharose CL-6B
↓
SP—Sephadex C-25
↓
Sephadex G-50
↓
Highly purified PTTH.

Each step of purification is described in detail in the extensive monograph of Ishizaki and Suzuki (1980).

The degree of purity of PTTH extracts can be determined using the bioassay introduced by Williams (1946) and developed by Ishizaki and Ichikawa (1967). The principle of this method consists of the injection of PTTH extracts into *Samia cynthia* larvae that have had the brain removed, within 20 h after pupation. Normally such pupae do not undergo metamorphosis to imagines after brain removal. The concentration of this hormone can be evaluated by determining the percentage of pupae initiating imaginal eclosion after injection of PTTH extract.

A new method of estimating PTTH activity in insect tissue extracts has been developed by Gibbs and Riddiford (1977). Larvae of *Manduca sexta* in the penultimate instar were ligated behind the head capsule just before PTTH release. Such ligation makes further development of the larva impossible. Injection of extract containing PTTH, however, allows continuation of development, to form headless larvae of the last instar or forms intermediate between the larva and the pupa. The extent of this response depends on the PTTH concentration in the tested extract.

The above method of determining PTTH concentration is for many reasons inconvenient, and has not found wide application. A new method has, however, been developed for investigating PTTH activity *in vitro* (Agui *et al.*, 1979; Bollenbacher *et al.*, 1979). It is the best method so far available, since it makes possible the study of the direct influence of this hormone on the prothoracic glands. This method became feasible after the development of extremely sensitive radioimmunological methods for the determination of ecdysone titres (see further text). The principle of the method consists in the evaluation of the rate of synthesis of ecdysone by the prothoracic glands *in vitro* before and after introduction of the extract containing PTTH.

The sensitivity of this method is confirmed by the fact that it can measure the prothoracicotropic activity of a single neurosecretory cell in the brain (Agui *et al.*, 1979).

3.7.2. Extraction of juvenile hormone (JH)

To determine the JH titre in insect haemolymph or in particular tissues, the hormone has to be extracted and at least partly purified. Extraction from haemolymph or tissues is with ether or ether-ethanol.

After extraction, all three ether fractions should be combined in one vessel and rapidly (within 15 min) evaporated on a water bath. The residue after evaporation is dissolved in a known volume of arachis oil and stored at low temperature until determination of its biological activity.

3.7.2.1. Extraction of JH from tissues

The method of extracting JH from insect tissues was developed by Schneiderman *et al.* (1965) and modified by de Wilde *et al.* (1971), and adapted for the estimation of the titre of the hormone in *Leptinotarsa decemlineata* tissues.

For extraction, take 10–15 beetles, weigh them, and after killing them add the same amount by weight of ether and then homogenize for 5 min. After homogenization leave the mixture to stand in a sealed vessel for 24 h. Then add 1–2 ml ether for further extraction. The mixture is passed through filter paper, and another 1–2 ml of ether is poured over the residue in the funnel, and the residue is left to stand for 24 h. The filtrates are mixed and the solvent is evaporated, while ether or pentane is poured over the residue. The solvent is again evaporated and the residue, after weighing, is dissolved in a known amount of arachis oil and stored in a refrigerator until testing.

3.7.2.2. Extraction of JH from haemolymph

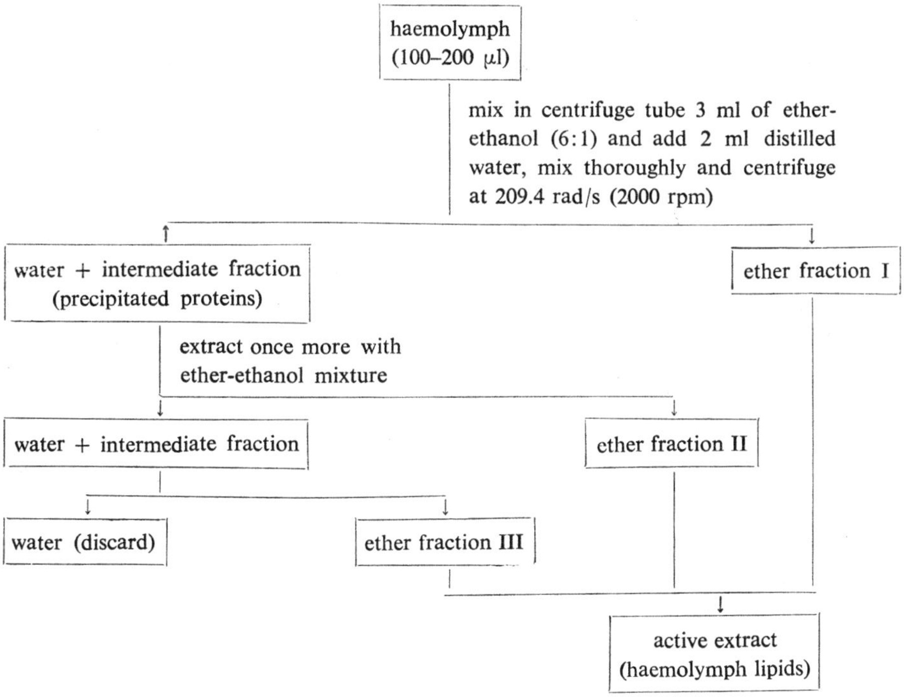

3.7.3. Methods of determining JH activity

The activity of extracts containing juvenile hormone can be determined using newly pupated *Galleria mellonella* individuals (Schneiderman *et al.*, 1965) or larvae on the first day after moulting to the last instar (Jarolim *et al.*, 1969).

3.7.3.1. JH assays in pupae

For this test a large number of *Galleria mellonella* pupae should be taken 24 h after pupation. There may be some difficulty in taking the pupae out of the cocoons, so the method developed by De Loof and Van de Veire (1972) for removing them is worth mentioning, since it saves time. To obtain a large number of pupae free from cocoons, the cocoons should be immersed for 15–30 s in a 1 N NaOH solution, then rapidly rinsed in water and dried on filter paper. The undissolved parts of the cocoon should then be removed. The mortality of the pupae obtained in this way is very low, usually not exceeding 5 per cent.

This method of cocoon removal is particularly effective when combined with the method devised by de Wilde *et al.* (1968); it is then possible to obtain pupae with very thin cocoons. For this purpose a large number of plastic tubes 3 cm long with inner diameter 4 cm are placed in the container in which the larvae are reared. The larvae readily crawl into these tubes and pupate there, spinning cocoons with very thin walls.

With an adequate number of pupae free of cocoons, the JH oil extract may be assayed by one of the methods described above. First a series of dilutions of the extract should be prepared using arachis oil as solvent. Then 12.5 mg of each of the dilutions is mixed with the same amount of wax. After thorough mixing (the wax may be slightly warmed) 1/20 of the total amount is placed on the pupa after a small fragment (0.25 mm^2) of cuticle and epidermis has been removed from the thorax. The wound is sealed with the mixture of wax (warmed to 50°C) and JH extract, and the pupae are placed in a thermostatically controlled chamber. The results are evaluated 6–7 days after application of the extract.

The appearance on the body of the moths that eclose from the experimental pupae of a fragment of brown cuticle, typical of pupae, indicates a positive result.

After calculation of the juvenile hormone concentration in the extract the "Galleria unit" is established. This is defined as the quantity of JH present in 0.625 mg of extract mixed with 0.625 mg of wax that results in preservation of the patches of pupal cuticle in the imago. To determine the JH concentration in the total amount of material extracted the following equation can be used:

Specific JH activity per gram fresh weight equals

$$\frac{1}{a} \cdot b \cdot \frac{c+d}{c} \cdot 1.6 \cdot 2^e \text{ Galleria units,}$$

where: a—total fresh weight (in g) used for extraction, b—weight (in mg) of ether extract obtained (after evaporation), c—amount of extract (in mg) placed on each pupa, d—amount of arachis oil (in mg) added, 2^e—final dilution of extract used in the assay.

The coefficient 1.6 is used since 1 Galleria unit is contained in 0.625 mg extract $\left(\frac{1000}{625} = 1.6\right)$.

Example of calculation of JH concentration:

Dissolve dry mass extracted with ether in 1 mg of arachis oil ($d = 1$) and prepare a series of dilutions at a 1:1 ratio.
For example:

1. 0.1 mg extract + 0.1 ml oil ($e = 1$),
2. 0.1 ml solution + 0.1 ml oil ($e = 2$),
3. 0.1 ml solution + 0.1 ml oil ($e = 3$) and so on.

Let us assume that, after performance of the test on pupae, dilution 3 corresponds to 1 Galleria unit. Thus, in order to find how many milligrams of extract correspond to this, the following values should be substituted in the calculation:

$a = 2.5$ g,
$b = 1200$ mg,
$c = 0.25$ mg,
$d = 100$ mg,
$e = 3$ (third dilution),
coefficient $= \frac{1000}{25} = 40.$

Specific activity of extract

$$\frac{1}{2.5} \cdot 1200 \text{ mg} \cdot \frac{0.25 \text{ mg} + 100 \text{ mg}}{0.25 \text{ mg}} \cdot 40 \cdot 2^3 =$$
$$= 157\,540 \text{ Galleria units per g fresh weight.}$$

3.7.3.2. Assays for extracts of JH and its analogues (JHA) in larvae

To check the biological activity of juvenile hormone extracts, and particularly that of synthetic JH analogues (juvenoids), the most commonly used method uses final instar larvae of the wax moth (Jarolim *et al.*, 1969). These authors adopted a five-point scale to score the morphological changes evoked by various concentrations of JH or JHA administered in acetone solution to the body surface of larvae. A similar five-point scale for morphological changes produced by JH was adopted by Riddiford (1972) for the silkworm *Hyalophora cecropia.* To each point on this scale there are corresponding characteristic morphological changes in wax moth larvae, as described in Table 1 (Plate 17).

Table 1

Scale of morphological changes caused by JH injected into *Galleria mellonella* larvae at various times of VII larval instar

Degree of morphological changes	Description of changes
5	superlarva (additional—VIII larval stage) morphologically not differing from normal larva, but has larger head capsule and body dimensions (Plate 17 A left)
4	morphologically similar to normal larva, has, however, longer antennae, slightly evaginated wing discs, and in some cases a fragment of pupal cuticle on the head and first body segment (Plate 17 A, right)
3	intermediate forms between larva and pupa with evaginated wing discs which reach a length of 5–7 mm. The larval antennae are elongated; pupal cuticle on the head and first and second segment of the thorax and the last abdominal segment (Plate 17 B)
2	intermediate form of pupal shape with larval cuticle on part of abdomen or limited only to several abdominal intersegmental membranes. Processes and wings are shorter than in pupa (Plate 17 C)
1	pupa with short wings (Plate 17 D, left)
0	normal pupa (Plate 17 D, right)

The highest biological activity on the five-point scale is attributed to JH or JHA solutions that induce additional larval stages, known as superlarvae. Sometimes the activity of the tested compound may be so high that not one, but two additional larval stages are obtained. In such a case greater dilutions should be prepared. JH and its analogues are usually diluted with acetone of high purity as solvent. The solution is pipetted from a calibrated capillary tube onto the body surface of the larva, anaesthetized with water, using 1 μl per individual. Control larvae receive the same amount of acetone. It was found that this compound does not induce morphological changes in developing *Galleria mello-*

nella larvae. For comparison of the activity of various JH extracts or juvenoids, the 50 per cent inhibitory dose (ID 50) is determined. The ID 50 is the amount of JH or its analogue in the tested material which inhibits metamorphosis in 50 per cent of larvae.

Morphological changes in the development of *Galleria mellonella* larvae similar to those described here can be obtained by implanting larvae with active *corpora allata* from final-instar larvae; this was done for the first time by Piepho (1939), whose results were confirmed by Sehnal (1968). It is therefore possible to assess the activity of *CA* of insects in various physiological states.

Instead of applying the JH or JHA extracts to the body surface, they can be injected into the body cavity. This is a more complicated procedure, but may be useful in investigations of the mechanism of action of this hormone.

Numerous other biological methods of determining the titre of juvenile hormone are given in a book by Slâma *et al.* (1974). He gives bioassay methods using larvae of *Tenebrio molitor*, *Rhodnius prolixus*, *Pyrrhocoris apterus*, *Drosophila melanogaster*, *Eurygaster intergriceps* (*Hemiptera*) and many others.

Mention should also be made of the sensitive biological test of juvenile hormone activity using the black mutant of *Manduca sexta* larvae devised by Truman *et al.* (1973). It is a very simple method, although rearing the larvae is troublesome.

All the biological methods of determining the juvenile hormone concentration in insect haemolymph and tissues described here have one essential drawback in that only the total JH pool can be determined; it is not possible to assay separately the four forms of the hormone (JH O, JH I, JH II and JH III). When it became possible to produce antibodies against the different kinds of JH (Lauer *et al.*, 1974a, b), many sensitive radioimmunological methods of determining JH in the insect body were developed. Two such methods have been published by Baehr *et al.* (1976) and Granger *et al.* (1979). The sensitivity of Granger's method reaches 50 pg JHI in the tested sample.

The rapid development of analytical techniques in recent years has been exploited in the development of sensitive methods of determining juvenile hormone titres at different stages of insect development. Gas chromatography and mass spectrometry are widely used in these assays (Mauchamp *et al.*, 1979; Rembold *et al.*, 1980).

3.7.4. Methods of extracting ecdysone

Ecdysone, like juvenile hormone, has to be extracted from the insect tissue before determination of its activity. The two methods usually employed involve extraction with water and with methanol. Both methods were developed by Karlson and Shaaya (1964).

3.7.4.1. Extraction of ecdysone with water

Aqueous ecdysone extracts of insect tissue are prepared as follows:

15 g insect tissue

homogenize in 50 ml H_2O, centrifuge for 20 min at 5000 rph

→ sediment | supernatant I

sediment: homogenize in 40 ml H_2O, centrifuge

→ sediment (discard) | supernatant II

supernatant II + supernatant I: heat on water bath for 30 min at 80–90°C, centrifuge for 20 min

→ sediment (discard) | supernatant

supernatant: concentrate on evaporator to 5 ml and dilute again with water to 10 ml, centrifuge

→ sediment (discard) | supernatant (crude extract, 10 ml)

supernatant (crude extract, 10 ml): mix thoroughly with 5 ml water saturated with butanol; repeat 4 times

→ water solution (discard) | butanol solution (20 ml)

butanol solution (20 ml): stir thoroughly with 20 ml water; repeat twice

→ water solution | butanol solution I

water solution: extract with 5 ml water saturated with butanol

→ water solution (discard) | butanol solution II

butanol solution I: dry on evaporator, dissolve in suitable amount of water

→ active extract

For the extraction of ecdysone 15 g of tissue or of whole insect is usually sufficient. The tissue is homogenized in 50 ml distilled water at room temperature for 2 min in a Teflon homogenizer. The material is then centrifuged at 5000 rpm, yielding supernatant I. The sediment should be homogenized again in 40 ml distilled water and centrifuged to give supernatant II. Both supernatants are combined and heated on a water bath at 80–90°C for 30 min. After centrifugation (20 min) a supernatant is obtained, which should be concentrated to 5 ml in an evaporator. To the concentrated solution 5 ml distilled water is added, and the precipitated proteins are centrifuged off. In this way 10 ml of crude extract are obtained, which should be agitated four times with 5 ml water saturated with butanol, giving 20 ml of butanol solution. The water solution is discarded. The butanol fraction should be shaken twice with 20 ml distilled water. The aqueous fractions (two) obtained in this way are combined and extracted once more with 5 ml butanol. The butanol fractions are dried in an evaporator. The sediment is dissolved in 0.2 ml water (1 ml if the ecdysone content is high) and the biological activity of the extract can then be determined.

3.7.4.2. Extraction of ecdysone with methanol

Methanol extracts of ecdysone are prepared according to the following scheme (see p. 64).

To extract ecdysone with methanol, 15 g of tissue or of whole insect is homogenized for 2 min in 50 ml methanol, producing supernatant I. Homogenize the sediment again, extract with 60 per cent methanol and centrifuge to obtain supernatant II. The two supernatants are combined and concentrated on an evaporator. The sediment is dissolved in 10 ml water. The aqueous solution (10 ml) is extracted four times with 5 ml water saturated with butanol. The combined butanol solutions should be shaken twice in 20 ml water to obtain butanol fraction I. The water fraction is extracted again with 5 ml butanol to give butanol fraction II and a water fraction which is discarded. The two butanol fractions are combined and 5 ml ether is added. The mixture obtained is extracted three times with 5 ml of 60% methanol, and after shaking it is centrifuged for 10 min at 3000 rpm. The lower layers should be pipetted off and combined. The combined water-methanol fractions are evaporated and the residue is dissolved in a known amount of water. The extract obtained can be used for determination of biological activity.

3.7.5. Determination of ecdysteroid concentration in insect tissue extracts

To determine ecdysone activity in insect tissue extracts a bioassay is usually run using ligated *Calliphora erythrocephala* larvae (Karlson and Shaaya, 1964).

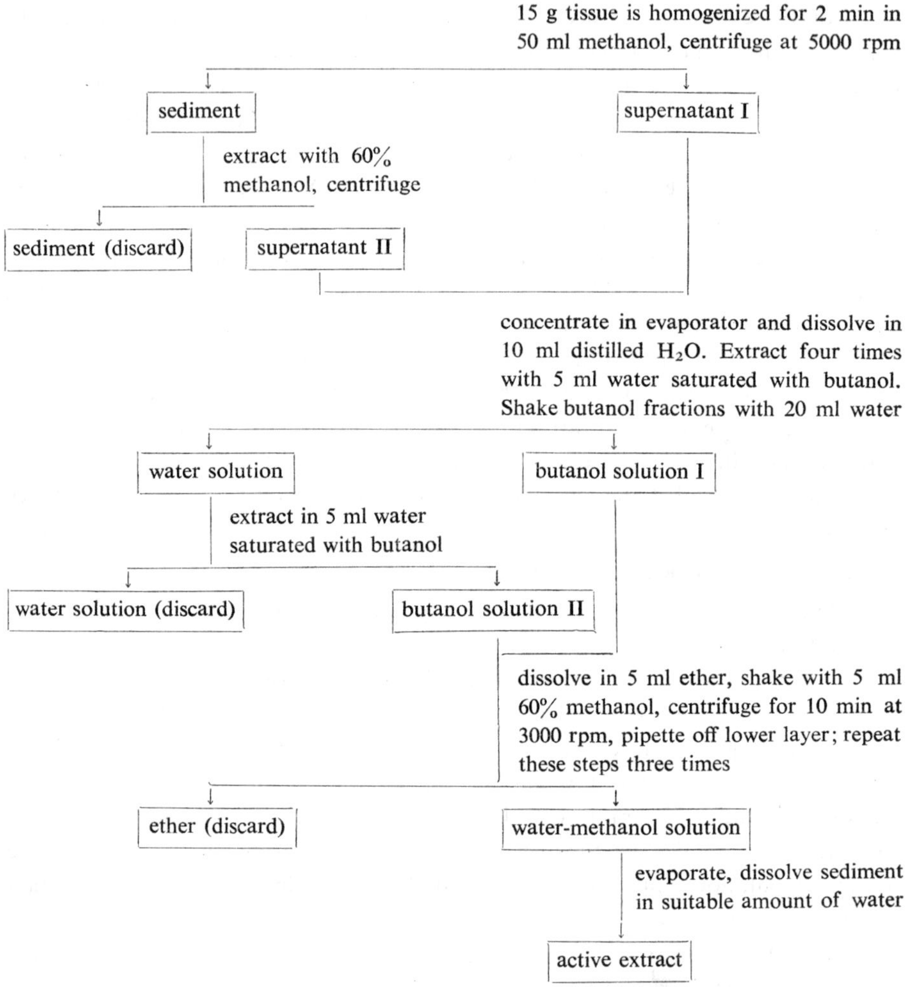

A much simpler, but no less sensitive, test is the bioassay using ligated larvae of the house fly *Musca domestica* (Kaplanis *et al.*, 1966; Adelung and Karlson, 1969). The attraction of this method lies in the ease with which fly larvae can be reared and the possibility of producing a large amount of material in a short time.

The tests are carried out with a group of 50–150 final-instar larvae that have emerged from the material upon which they have been feeding. They are ligated between the 7th and 8th body segment. The ligature should be drawn tight, but should allow the larva to advance and retract its mouth parts. After fitting the ligature, the larvae are transferred to Petri dishes lined with filter paper to prevent the larvae from dehydrating when placed in a thermostatically controlled chamber at 28°C. Twelve hours after ligation, by which time the anterior part will have pupated, the larvae are transferred to separate dishes. After 5–6 h they can be injected with the ecdysone extract to be assayed.

The extract prepared as described above is injected (3 μl) into the abdomen. This is usually done with a microsyringe. The needle is moistened with 0.1 per cent sodium perchloride, introduced posterior to the ligature and inserted almost to the end of the abdomen. The second ligature is placed on the third segment posterior to the first one. After injection, the second ligature is tightened to prevent leakage of haemolymph.

The control group consists of larvae injected with a known quantity of ecdysone in 3 μl distilled water.

The result is evaluated 24 h after injection by comparison with the control. The scores allocated are 1 point for complete pupation, 0.75 for partial pupation, 0.5 for the appearance of a small amount of pupal cuticle, 0.25 for a trace of pupal cuticle, and O when there are no traces of pupal cuticle (Fig. 3.3). The results can also be expressed as the percentage of larvae surviving after injection of ecdysone extract. Ten ligated abdomens are usually used for each dilution of the tested sample. The method is highly sensitive, within the limits of 1–100 nanograms ecdysone per abdomen.

In addition to the methods of determining the ecdysone content in biological material described here, Sláma *et al.* (1974) give several other methods. These refer mainly to *Diptera* and *Lepidoptera*, however, and are not commonly used. The problem lies mainly in the difficulty of finding a stage of the insect that is highly sensitive to ecdysone. It seems, therefore, that the most suitable subjects for testing ecdysone activity in biological material are diapausing pupae with the brain removed (Williams, 1968).

Injection of these larvae with ecdysone-containing extract sooner or later (depending on ecdysone concentration) interrupts diapause. This method, however, is time-consuming and is not widely used.

As for JH, radioimmunological and chemical methods find increasingly wide application. These methods, of course, became possible with the availability of ecdysone antibodies (Lauer *et al.*, 1974).

The radioimmunological methods (RIA) developed by Borst and O'Connor (1974) and Horn *et al.* (1976) are relatively simple. The use of RIA to determine ecdysone and 20-hydroxyecdysone concentrations must be preceded by extraction

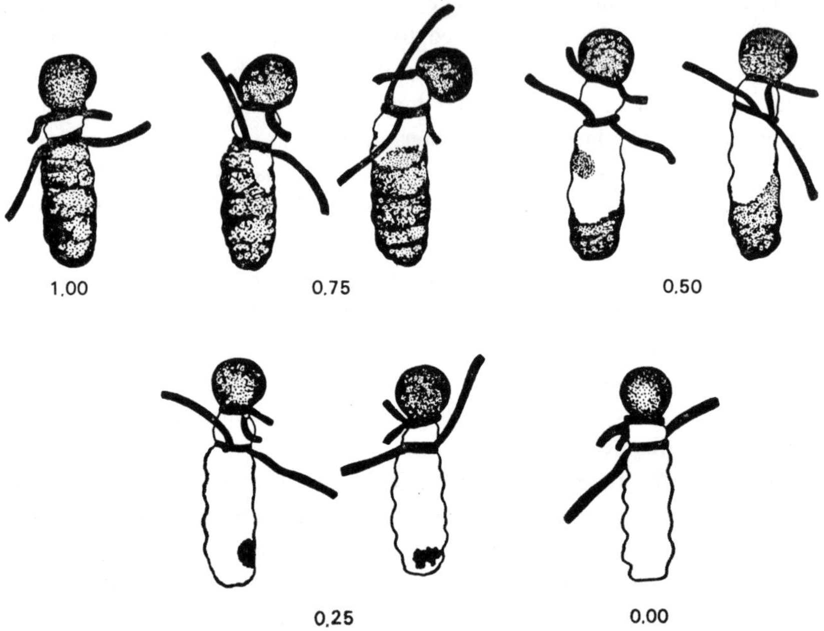

Figure 3.3. Evaluation of ecdysteroid activity according to the degree of pupation of ligated house fly larvae. Negative result (score 0) corresponds to the complete of pupal cuticle; scores 0.25 to 1 correspond to increasing proportions of pupal cuticle. After Adelung and Karlson (1969).

of the hormones from the biological material with methanol and purification by thin-layer chromatography. It should be added that the radioimmunological method is about 5–7 times more sensitive to 20–hydroxyecdysone than to ecdysone. Chang and O'Connor (1979) give a detailed analysis of the immunological methods, pointing out their unquestionable advantages but also calling attention to their shortcomings. The reader will also find in their paper detailed recommendations as to the use of these methods.

Less time-consuming analytical methods for ecdysone, but requiring intricate apparatus, are high-pressure liquid chromatography (HPLC) and gas chromatography (Gande and Morgan 1979; Lafont and Martin-Somme 1979).

3.7.6. Simultaneous determination of JH and ecdysone titres

It is frequently necessary in endocrinological investigations on insects to find the relative proportions of JH and ecdysone in the tissues and haemolymph.

In such studies it is obviously necessary to determine the concentrations of both these hormones in the same sample. To meet these requirements Hsiao and Hsiao (1977) developed a method of extracting both these hormones from insect tissues. About 120 g of fresh tissue is freeze-dried and homogenized in 50 ml of anhydrous methanol. The sample thus prepared is stored at 4°C for about 12 h and then centrifuged for 10 min at 5000 rpm. The sediment is again washed with 50 ml methanol, but at room temperature. The residue is extracted once more, this time for 30 min in boiling methanol. All methanol extracts are evaporated under reduced pressure at 50°C. The sediment is fractionated in 15 ml methanol plus water (1:1 V/V) and 15 ml hexane. Separation is achieved by centrifugation at 3000 rpm for 3 min.

The hexane fraction contains JH, and after evaporation and suitable preparation may be used for determining the activity of this hormone (see Section 3.7.3). The methanol-water fraction containing ecdysone is evaporated under reduced pressure at 60°C. The sediment is washed with butanol and again evaporated. The ecdysone content can be determined in the sediment thus prepared (see Section 3.7.5).

3.7.7. Determination of bursicon activity

Bursicon—the hormone controlling the processes of hardening and darkening of the cuticle of imagines after eclosion—is released into the haemolymph by the abdominal ganglia of the nervous system. Vincent (1972) devised a bioassay for determining the concentration of this hormone using newly emerged imagines of *Phormia terraenovae*, based on the results of Frankel and Hsiao (1965).

3.7.7.1. Preparation of homogenates for bursicon determination

Homogenates for bursicon determination are usually obtained by homogenization of tissue in physiological saline containing 7.5 g NaCl and 0.375 g KCl/l distilled water. Depending on the amount of tissue, different amounts of liquid are used for homogenization. For instance, two ganglia of the ventral nerve cord are homogenized in 50 μl liquid, and *corpora cardiaca* from one insect in 20 μl. The effectiveness of homogenization can be estimated from the degree of clarity of the homogenate.

Insect haemolymph can be used for determination of bursicon concentration after dilution with physiological saline.

3.7.7.2. Bioassay for bursicon activity

To obtain an adequate quantity of material necessary for the assay, pupae of *Phormia terraenovae* are collected over a long period and kept in a refrigerator to inhibit further development. The pupae are transferred to an incubator, and immediately after eclosion of the imagines, whilst the insects are still white, a ligature is fitted just behind the head segment. The ligated insects are kept for 5 h, and any individuals in which the cuticle has then become coloured are discarded.

Each experimental group requires at least five ligated insects, which are placed in Petri dishes lined with moist filter paper. Then each insect is injected with 5 μl of the solution to be tested through a glass capillary tube. The capillary tube is introduced just under the cuticle into the anterior part of the body. This method of injection prevents leakage of haemolymph. Insects in which leakage of haemolymph is noted should be discarded. The results are evaluated 4–6 h after injection, using the seven point scale shown in Table 2.

Table 2
Scale of changes in cuticle colouring caused by injection of extracts containing bursicon

Score	Characteristic of changes
0	no colouring
1	trace colouring occupying less than 30% of dorsal abdominal surface
2	colouring in the form of spots on the thorax or abdomen occupying more than 30% of dorsal surface of abdomen
2.5	abdomen completely dark, thorax not coloured or with traces of colour
3	thorax and abdomen dark or almost dark
3.5	mottled colouring on end of abdomen
4	completely dark insect

The scores for five insects of each group are summed (maximal number of points 20) and the result is multiplied by 5 to obtain the test result. The maximal possible value is 100. The experimental result is expressed as a percentage of the theoretical maximal value.

This test for bursicon activity gives reproducible results. If there are any differences they are usually due to unprecise injection of the test substance.

Descriptions of several other methods of determining bursicon activity in insect tissues are given in the extensive paper of Seligman (1980). He gives an account of methods based on the colouring of *Schistocerca gregaria* larvae, *Tenebrio molitor* pupae and wings of freshly eclosed *Manduca sexta* imagines.

3.8. Investigation of eclosion hormone activity

It was shown by Truman (1971) that eclosion of the mature insect from the pupa, associated with a whole chain of characteristic changes in the behaviour of the pupa, is initiated by the appearance of the eclosion hormone in the haemolymph. A very sensitive bioassay has been developed for determining the activity of this hormone in *Antheraea pernyi* pupae (Truman, 1973). The method is based on the stimulation of premature moulting by injection of extracts containing the eclosion hormone. The extract is injected about 10 h before eclosion is expected. If the injected material contains the hormone, moulting will occur 1.5 to 4.5 h after treatment. If moulting does not occur the result is considered negative. A score can be established describing the activity of the hormone in the sample, based on observations made 1.5 to 4.5 h after injection.

Several other biological methods of determining eclosion hormone activity have been developed (Reynolds and Truman, 1980). One interesting method involves the characteristic movements of the ligated abdomen of the *Hyalophora cecropia* pupa just before eclosion of the imago. Such movements can be evoked by injecting extract containing eclosion hormone, and this is the basis of a very sensitive bioassay (Reynolds, 1977).

3.9. Determination of activity of other insect hormones

So far only the methods that are commonly used in insect endocrinology have been described. There are many new methods, however, which have not yet found their way into biological laboratories. These methods mainly concern poorly known hormones, and hormones that are not present in all insect species. One example of a hormone of limited occurrence is the diapause hormone detected to date only in the silkworm *Bombyx mori*. An assay for this hormone is described in detail by Isobe and Goto (1980).

One relatively little known hormone is proctolin, isolated from the gut of *Periplaneta americana*, which stimulates contractions of the insect's gut. It is a pentapeptide of known composition; its biological activity can be determined by the simple method of Starratt and Steele (1980).

Other poorly known hormones include the diuretic hormone, the concentration of which can be evaluated by a special bioassay using *Rhodnius prolixus* (Maddrell and Gardiner, 1976), and insect hormones similar in action to insulin and glucagon in vertebrates. Their activity can be determined by radioimmunological methods (Kramer, 1980).

3.10. Summary

The seclection of methods for investigating insect endocrinology presented here is not exhaustive. The aim has been to indicate to the reader the wide range of possibilities that exist for studying the functions of the insect endocrine system. The important sphere of research on the action of hormones at the molecular level has been purposely omitted, since a separate chapter of this book is devoted to these problems. Only those methods are presented which are commonly used at present in insect endocrinology.

Some methods have been given no more than a mention, such as the pharmacological methods, because they are still being refined and improved. They will probably be very useful, however, in future investigations of insect hormones.

3.11. References

Adelung D., Karlson P. 1969. Eine verbesserte, sehr empfindliche Methode zur biologischen Auswertung des Insektenhormones Ecdyson. *J. Insect Physiol.*, **15**; 1301–1307.

Agui N., Grenger N. A., Gilbert L. I., Bollenbacher W. E. 1979. Cellular localization of the insect prothoracicotropic hormone. *In vitro* assay of a single neurosecretory cell. *Proc. Nat. Acad. Sci. USA*, **76**; 5694–5698.

Baehr J. C., Pradelles P. H., Lebreux C., Cassier P., Dray F. 1976. A simple and sensitive radioimmunoassay of insect juvenile hormone using an iodinated tracer. *FEBS Letters*, **69**; 123–128.

Bollenbacher W. E., Agui N., Granger N. A., Gilbert L. I. 1979. *In vitro* activation of insect prothoracic glands by the prothoracicotropic hormone. *Proc. Nat. Acad. Sci.* **76**; 5148–5152.

Borst D. W., O'Connor J. D. 1974. Arthropod molting hormone: Radioimmunoassay. *Science*, **178**; 418–419.

Brooks M. A. 1957. Growth retarding effect of carbon dioxide anesthesia on the German cockroach. *J. Insect Physiol.*, **1**; 76–84.

Cameron M. L., Steele J. E. 1959. Simplified aldehyde fuchsin staining of neurosecretory cells. *Stain Technol.*, **34**; 265–266.

Chang E. S., O'Connor J. O. 1979. Arthropod molting hormones. [In:] *Methods of Hormone Radioimmunoassay*, pp. 797–814. Academic Press, New York, San Francisco, London.

Cymborowski B. 1968. Badania nad neurohormonalną regulacją dobowego cyklu aktywności ruchowej świerszcza domowego (*Acheta domesticus* L.). Praca doktorska [Investigations on neurohormonal regulation of diurnal motor activity of the cricket (*Acheta domesticus* L.)]. Ph.D. thesis, Univ. of Warsaw.

Cymborowski B., Dutkowski A. 1969. Circadian changes in RNA synthesis in the neurosecretory cells of the brain and suboesophageal ganglion of the house cricket. *J. Insect Physiol.*, **15**; 1187–1197.

Cymborowski B. 1970. The assumed participation of 5-hydroxytryptamine in regulation of the circadian rhythm of locomotor activity in *Acheta domesticus* L., *Comp. gen. Pharmac.*, **1**; 316–322.

Cymborowski B., Dutkowski A. 1970. Circadian changes in protein synthesis in the neurosecretory cells of the central nervous system of *Acheta domesticus. J. Insect Physiol.*, **16**; 341–348.

Cymborowski B. 1970a. Investigations on the neurohormonal factors controlling circadian rhythm of locomotor activity in the house cricket (*Acheta domesticus* L.). I. The role of the brain and suboesophageal ganglion. *Zool. Pol.*, **20**; 103–125.

Cymborowski B. 1970b. Investigations on the neurohormonal factors controlling circadian rhythm of locomotor activity in the house cricket (*Acheta domesticus* L.). II. Daily histochemical changes in the neurosecretory cells of the *pars intercerebralis* and suboesophageal ganglion. *Zool. Pol.*, **20**; 127–149.

Cymborowski B. 1972. The effect of reserpine on locomotor activity of crickets. *Exp. Physiol. Biochem.*, **5**; 247–255.

Cymborowski B., Brady J. 1972. Insect circadian rhythm transmitted by parabiosis—a re-examination. *Nature New Biology*, **296**; 221–222.

Cymborowski B. 1973a. Control of the circadian rhythm of locomotor activity in the house cricket. *J. Insect Physiol.*, **19**; 1423–1440.

Cymborowski B. 1973b. Les variations diurnes de l'activité des cellules neurosécrétrices du cerveau du grillon (*Acheta domesticus* L.). *Acrida*, **2**; 299–306.

Cymborowski B., Kryspin I. M. 1975. Effect of reserpine on development and its neuro-endocrine regulation in *Galleria mellonella*. *Gen. Pharmac.*, **6**; 209–213.

Cymborowski B. 1976. *Zegary biologiczne* [Biological clocks]. PWN, Warszawa.

Day M. F., Grace T. D. C. 1959. Culture of insect tissues. *Ann. Rev. Entomol.*, **4**; 17–38.

De Loof A., Van de Veire M. 1972. Time saving improvements in the *Galleria* bioassay for juvenile hormone. *Experientia*, **28**; 366–367.

Doane W. W. 1961. Development physiology of the mutant female adipose tissue of *Drosophila melanogaster*—III. *Corpus allatum* complex and ovarian transplantations. *J. exp. Zool.*, **146**; 275–298.

Dogra G. S., Tandan B. K. 1964. Adaptation of certain histological techniques for *in situ* demonstration of the neuroendocrine system of insects and other animals. *Quart. J. micr. Sci.*, **105**; 455–466.

Dutkowski A. B., Cymborowski B. 1973. The role of the suboesophageal ganglion in regulation of RNA synthesis in some tissues. *J. Insect. Physiol.*, **19**; 1533–1540.

Edwards L. J., Patton R. L. 1965. Effect of carbon dioxide anesthesia on the house cricket, *Acheta domesticus* (*Orthoptera*: *Gryllidae*). *Ann. Entomol. Soc. Amer.*, **58**; 828–832.

Frankel G., Hsiao C. 1965. Bursicon, a hormone which mediates tanning of the cuticle in the fly and other insects. *J. Insect Physiol.*, **11**; 513–566.

Gande A. R., Morgan E. D. 1979. Ecdysteroids in the developing eggs of the desert locust, *Schistocerca gregaria*. *J. Insect Physiol.*, **25**; 289–293.

Gibbs D., Riddiford L. M. 1977. Prothoracicotropic hormone in *Manduca sexta*. Localization by a larval assay. *J. exp. Biol.*, **66**; 255–266.

Gomori G. 1950. Aldehyde-fuchsin: a new stain for elastic tissue. *Amer. J. Clin. Pathol.*, **20**; 665–666.

Granger N. A., Bollenbacher W. E., Vince R., Gilbert L. I., Baehr J. C., Dray F. 1979. *In vitro* biosynthesis of juvenile hormone by the larval *corpora allata* of *Manduca sexta*. Quantification by radioimmunoassay. *Molecular and Cellular Endocrinology*, **16**; 1–17.

Halmi N. S. 1952. Differentiation of two types of basophils in the adenohypophysis of the rat and the mouse. *Stain Technol.*, **27**; 61–64.

Hang F-M. S. 1973. *Heavy Metals in the Brain*. Springer-Verlag, Berlin.

Highnam C. K. 1961. Induced changes in the amounts of material in the neurosecretory system of the desert locust. *Nature*, **191**; 199–200.

Horn D. H. S., Wilkie J. S., Sage B. A., O'Connor J. D. V. 1967. A high activity antiserum specific for the ecdysone nucleus. *J. Insect Physiol.*, **22**; 901–905.

Hsiao T. H., Hsiao C. 1977. Simultaneous determination of molting and juvenile hormone titers of the Greater wax moth. *J. Insect Physiol.*, **23**; 89–93.

Ishizaki H., Ichikawa M. 1967. Purification of the brain hormone of the silkworm, *Bombyx mori*. *Biol. Bull.*, **133**; 355–368.

Ishizaki H., Suzuki A. 1980. Prothoracicotropic hormone. [In:] *Neurohormonal Techniques in Insects*. T. A. Miller (ed.), pp. 244–276. Springer-Verlag, New York, Heidelberg, Berlin.

Isobe M., Goto T. 1980. Diapause hormones. [In:] *Neurohormonal Techniques in Insects*. T. A. Miller (ed.), pp. 216–243. Springer-Verlag, New York, Heidelberg, Berlin.

Isobe M., Goto T. 1980. Diapause hormones. [In:] *Neurohormonal Techniques in Insects* T. A. (ed.), pp. 216–243. Springer-Verlag, New York, Heidelberg, Berlin.

Jacklet J. W., Cohen M. J. 1967. Nerve regeneration: Correlation of electrical, histological and behavioral events. *Science*, **156**; 1640–1643.

Jarolim V., Hejno K., Sehnal F., Šorm F. 1969. Natural and synthetic materials with insect hormone activity. 8. Juvenile activity of the farnesane-type compounds in *Galleria mellonella*. *Life Sci.*, **8**; 831–841.

Kaplanis J. N., Tabor L. A., Thompson M. J., Robbins W. E., Shortino T. I. 1966. Assay for ecdysone (molting hormone) activity using the house fly, *Musca domestica* L. *Steroids*, **8**; 625–631.

Karlson P., Shaaya E. 1964. Der Ecdysontiter während der Insekten-entwicklung—I. Eine Methode zur Bestimmung des Ecdysongehalts. *J. Insect Physiol.*, **10**; 797–801.

Kiguchi K., Riddiford L. M. 1978. A role of juvenile hormone in the pupal development of the tobacco hornworm, *Manduca sexta*. *J. Insect Physiol.*, **24**; 673–680.

Kopeć S. 1917. Experiments on metamorphosis of insects. *Bull. int. Acad. Sci. Cracovie* (*B*), 57–60.

Kramer K. J. 1980. Insulin-like and glucagon-like hormones in insects. [In:] *Neurohormonal*

Techniques in Insects. T. A. Miller (ed.), pp. 116–136. Springer-Verlag, New York, Heidelberg, Berlin.

Lafont R., Martin-Somme G. 1979. Separation of ecdysteroids by using high-pressure liquid chromatography on microparticulate supports. *J. Chromatogr.*, **170**; 185–194.

Lauer R. C., Solomon P. H., Nakanishi K., Erlanger B. F. 1974a. Antibodies to the insect moulting hormone β-ecdysone. *Experientia*, **30**; 560–562.

Lauer R. C., Solomon P. H., Nakanishi K., Erlanger B. F. 1974b. Antibodies to insect C_{16}-juvenile hormone. *Experientia*, **30**; 558–560.

Maddrell S. H. P., Gardiner B. O. C. 1976. Diuretic hormone in adult *Rhodnius*: total store and speed of release. *Physiol. Entomol.*, **1**; 265–269.

Mahon D. C., Nair K. K. 1975. A comparison of aldehyde fuchsin and alcian blue staining of neurosecretory material in *Oncopeltus fasciatus*. *Cell Tiss. Res.*, **161**; 477–484.

Marks E. P. 1971. Cultivation of insect endocrine glands *in vitro*. *Current Topics in Microbiology and Immunology*, **55**; 75–85.

Marks E. P., Holman G. M. 1974. Release from the brain and acquisition by the *corpus cardiacum* of a neurohormone *in vitro*. *J. Insect Physiol.*, **20**; 2087–2093.

Marks E. P., Ittycheriah P. I., Leloup A. M. 1972. The effect of β-ecdysone on insect neurosecretion *in vitro*. *J. Insect Physiol.*, **18**; 847–850.

Masner D., Huot L., Corrivault G. W., Prudhomme J. C. 1970. Effect of reserpine on the function of the gonads and its neuroendocrine regulation in tenebrionid beetles. *J. Insect Physiol.*, **16**; 2327–2344.

Martignoni M. E. 1962. Insect tissue culture: A tool for the physiologist. [In:] *Insect Physiology*. V. J. Brookes (ed.), pp. 89–110. Oregon State University Press, Oregon.

Mauchamp B., Lafont R., Hardy M., Jourdain D. 1979. Analysis of insect juvenile hormones by gas chromatography mass spectrometry: Problems of sample preparation and choice of detection procedure, *Biomed. Mass Spectr.*, **6**; 276–281.

Oberlander H., Berry S. J., Krishnakumaran A., Schneiderman H. 1965. RNA and DNA synthesis during activation and secretion of the prothoracic glands of saturniid moths. *J. Exp. Zool.*, **159**; 15–32.

Paget G. S. 1959. Aldehyde-thionin: a stain having similar properties to aldehyde-fuchsin. *Stain Technol.*, **34**; 223–226, 1959.

Patton R. L., Edwards L. J., Gilmore S. K. 1968. Delivering safe levels of CO_2 for insect anesthesia. *Ann. Entomol. Soc. Amer.*, **61**; 1046–1047.

Pearse A. G. K. 1960. Histochemistry Theoretical and Applied. Churchill, London.

Piepho H. 1939. Hemmung der Verpuppung durch *Corpora allata* von Jungraupen bei der Wachsmotte *Galleria mellonella* L. *Naturwiss.*, **27**; 675–676.

Piepho H. 1940. Über die Hemmung der Verpuppung durch *Corpora allata* Untersuchungen an der Wachsmotte *Galleria mellonella* L. *Biol. Zbl.*, **60**; 367–393.

Pihl E., Falkner S. 1967. Trials to modify the sulphide silver method for ultrastructural tissue localization of heavy metals. *Acta Histochem.* (Jena), **27**; 34–41.

Rembold H., Hagenguth H., Rascher J. 1980. A sensitive method for detection and estimation of juvenile hormones from biological samples by glass capillary combined gas chromatography—selected ion monitoring mass spectrometry. *Analytical Biochem.*, **101**; 356–363.

Reynolds S. E. 1977. Control of cuticle extensibility in the wings of adult *Manduca* at the time of eclosion: effects of eclosion hormones and bursicon. *J. Exp. Biol.*, **70**; 27–39.

Reynolds S. E., Truman J. W. 1980. Eclosion hormones. [In:] *Neurohormonal Techniques in Insects*. T. A. Miller (ed.), pp. 196–215. Springer-Verlag, New York, Heidelberg, Berlin.

Riddiford L. M. 1972. Juvenile hormone in relation of the larval-pupal transformation of the *Cecropia* silkworm. *Biol. Bull.*, **142**; 310–325.

Rogers A. W. 1973. *Techniques of Autoradiography*. Elsevier Scientific Publishing Company, Amsterdam, London, New York.

Sawicki W. 1963. Autoradiografia w badaniach cytochemicznych [Autoradiography in cytochemical investigations]. *Post. Biochem.*, **9**; 257–276.

Scharrer B., Scharrer E. 1944. Neurosecretion. VI. A comparison between the *intercerebralis-cardiacum-allatum* system of the insects and the hypothalamo-hypophyseal system of the vertebrates. *Biol. Bull.*, **87**; 242–251.

Schneiderman H. A., Krishnakumaran A., Kulkami V. G., Friedman L. 1965. Juvenile hormone activity of structurally unrelated compounds. *J. Insect Physiol.*, **11**; 1641–1649.

Schoumacker H., van Damme N. 1971. Maillet's OsO_4—ZnI_2 fixative and alcian blue staining in the study of neurosecretion: invertebrates. *Stain Technol.*, **46**; 233–327.

Sehnal F. 1968. Influence of the *corpus allatum* on the development of internal organs in *Galleria mellonella* L. *J. Insect Physiol.*, **14**; 73–85.

Sehnal F., Granger N. A. 1975. Control of *corpora allata* function in larvae of *Galleria mellonella*. *Biol. Bull.*, **148**; 106–116.

Seshan K. R. 1976. *In vitro* analysis of the insect neuroendocrine organs. Entomon, **1**; 7–16.

Seligman I. M. 1980. Bursicon. [In:] *Neurohormonal Techniques in Insects*. T. A. Miller (ed.), pp. 137–153. Springer-Verlag, New York, Heidelberg, Berlin.

Sláma K. 1964. Hormonal control of respiratory metabolism during growth, reproduction and diapause in female adults of *Pyrrhocoris apterus* L. (*Hemiptera*). J. *Insect Physiol.*, **10**; 283–303.

Sláma K., Romaňuk M., Šorm F. 1974. Insect hormones and bioanalogues. Springer-Verlag, Wien, New York.

Starratt A. N., Steele R. W. 1980. Proctolin: bioassay, isolation and structure. [In:] *Neurohormonal Techniques in Insects*. T. A. Miller (ed.), pp. 1–30. Springer-Verlag, New York, Heidelberg, Berlin.

Strong L. 1963. A simple apparatus for use in removing *corpora allata* from locusts. *Bull. entomol. Res.*, **54**; 19–21.

Suzuki A., Isogai A., Horii T., Ishizaki H., Tamura S. 1975. A simple procedure for partial purification of Silkworm brain hormone. *Agr. Biol. Chem.*, **39**; 2157–2162.

Truman J. W. 1971. Physiology of insect ecdysis. The eclosion behaviour of saturniid moths and its hormonal release. *J. Exp. Biol.*, **54**; 805–814.

Truman J. W. 1973. Physiology of insect ecdysis. II. The assay and occurrence of the eclosion hormone in the Chinese Oak Silkmoth, *Antheraea pernyi*. *Biol. Bull.*, **144**; 200–211.

Truman J. W., Riddiford L. M., Safranek L. 1973. Hormonal control of cuticle coloration in the tobacco hornworm, *Manduca sexta*: basis of an ultrasensitive assay for juvenile hormone, *J. Insect Physiol.*, **19**; 195–203.

Truman J. W., Reiss S. E. 1976. Dendritic reorganization of an identified motoneuron during metamorphosis of the tobacco hornworm moth. *Science*, **192**; 477–479.

Tyrer N. M., Bell E. M. 1974. The intensification of cobalt-filled neurone profiles using a modification of Timm's sulphide-silver method. *Brain Res.*, **73**; 151–155.

Wigglesworth V. B. 1952a. Hormone balance and the control of metamorphosis in *Rhodnius prolixus* (*Hemiptera*). *J. Exp. Biol.*, **29**; 620–631.

Wigglesworth V. B. 1952b. The thoracic gland in *Rhodnius prolixus* (*Hemiptera*) and role in moulting. *J. Exp. Biol.*, **29**; 561–570.

Wigglesworth V. B. 1955. The breakdown of the thoracic gland in the adult insect, *Rhodnius prolixus*. *J. Exp. Biol.*, **32**; 485–491.

Wigglesworth V. B. 1957. The action of growth hormones in insects. *Symp. Soc. exp. Biol.*, **11**; 204–227.

Wigglesworth V. B. 1958. Some methods for assaying extracts of the juvenile hormone in insects. *J. Insects Physiol.*, **2**; 73–84.

Wilde J., de, Staal G. B., De Kort C. A. D., De Loof A., Baard G. 1968. Juvenile hormone titer in the haemolymph as a function of photoperiodic treatment in the adult Colorado beetle, *Leptinotarsa decemlineata* Say. *Konikl. Nederl. Acad. Wettenschap.*, **71**; 312–326.

Wilde J., de, Staal G. B., Oostra R. 1971. Reproduction, diapause and the juvenile hormone titre in the adult female *Leptinotarsa decemlineata* (Say.). [In:] *Insect Endocrines*, pp. 45–49. *Supl. Acta entomol. Bohemoslov.*

Williams C. M. 1946. Physiology of insect diapause, the role of the brain in the production and termination of pupal dormancy in the giant silkworm, *Platysamia cecropia*. *Biol. Bull.*, **90**; 234–243.

Williams C. M. 1946. Continuous anaesthesia for insects. *Science*, 103; 57.

Williams C. M. 1968. Ecdysone and ecdysone-analogue: Their assay and action on diapausing pupae of the *Cynthia* silkworm. *Biol. Bull. Woods Hole*, **134**; 344–355.

Vincent J. F. V. 1972. The dynamics of release and possible identity of bursicon in *Locusta migratoria migratorioides*. *J. Insect Physiol.*, **18**; 757–780.

Yamazaki M., Kobayashi M. 1969. Purification of the proteinic brain hormone of the silkworm, *Bombyx mori*. *J. Insect Physiol.*, **15**; 1981–1990.

Chapter 4

Hormonal regulation of insect growth and metamorphosis

The phenomena of insect development and metamorphosis have long interested scientists. The pioneering work of Kopeć, to whose memory the International Conference held in 1980 was devoted (Sehnal *et al.*, 1981), demonstrated in 1912–1917 that hormones undoubtedly participate in these processes. Further intensive research fully confirmed the results of Kopeć, and the development and metamorphosis of insects became the object of studies on growth regulation and cell differentiation. Insects are a particularly convenient object for such investigations, since these processes occur over extended periods, lasting throughout the life of the insect, whereas in other organisms they are usually limited to the embryonic period.

4.1. General characteristics of insect development

The most characteristic trait of insects is their external skeleton in the form of a hard, sclerotized cuticle, from which periodical emergence is necessary to allow growth and development. Throwing off of the old cuticle and formation of the new one occurs particularly during the period of larval growth after which metamorphosis takes place. In primitive insect species of the *Apterygota*, the differences between the larval and the adult forms are slight, being mainly concerned with the formation of new processes on the cuticle, and hairs on the legs and reproductive organs (Fig. 4.1). In insects of the *Exopterygota* changes are somewhat more pronounced. For example, in more advanced larval stages, wing discs appear which are transformed into the functional wings of the adult. The appearance of the reproductive organs also gradually changes. In the *Endopterygota*, the successive larval forms differ mainly in size, whereas the essential metamorphosis occurs during pupation and imaginal ecdysis. Here the pupa is a kind of transitional stage between the larva and the imago. In higher *Diptera* an almost complete lysis of the larval cells occurs in the pupal stage, while the imago

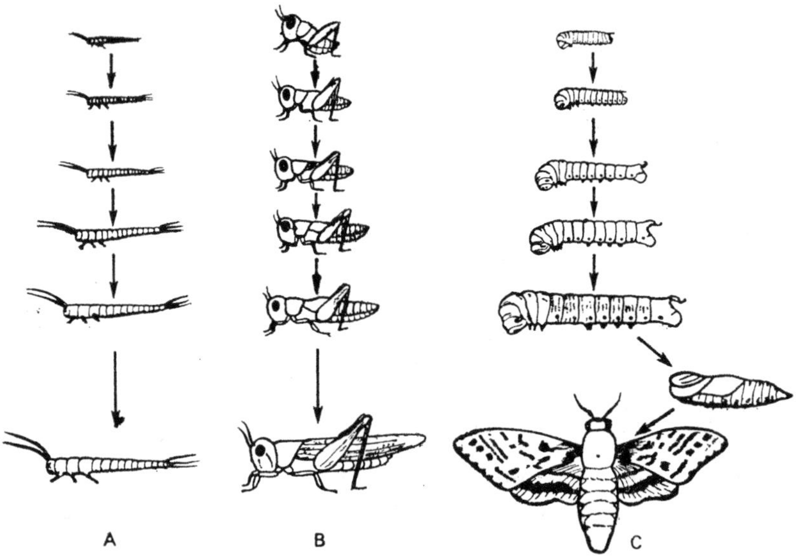

Figure 4.1. Diagram illustrating three main types of insect development and metamorphosis. After Highnam and Hills (1977), with permission.
A—*Apterygota*, B—*Exopterygota*, C—*Endopterygota*

arises from small groups of undifferentiated cells (imaginal discs and histoblasts) which exist in the larvae but do not develop. One might say that the larva is a kind of mosaic consisting of three main cell types:

1. typical larval cells which die towards the end of larval development (e.g. cells of the alimentary tract),
2. cells which transform to various pupal structures,
3. cells of the imaginal form which appear to be stored in the larva without noticeable changes in their structure and function.

How do particular cell types "know" the place and time of their development? Any attempt to answer this question requires a knowledge of the hormonal changes occurring during development and metamorphosis.

4.2 Hormones associated with development and metamorphosis

4.2.1. The prothoracicotropic hormone (PTTH)

Many investigators have attempted to obtain PTTH in pure form and determine its chemical composition. Noteworthy are the studies of Ishizaki and Ichikawa (1967) who extracted this neurohormone from 118 000 brains of silkworm

(*Bombyx mori*) pupae and purified it about 8000-fold. Further studies by Yamazaki and Kobayashi (1969) purified PTTH using about 100 000 larval brains of *Bombyx mori*. In both cases, however, electrophoretic analysis indicated the presence of certain impurities in the preparations obtained which prevented the accurate determination of the chemical structure of PTTH. At present the prevailing view seems to be that PTTH is a protein with a molecular weight of about 5000, consisting of at least two parts: activating factor I and activating factor II (Birkenbeil and Gersch, 1974). This substance is soluble in water and is thermostable—it is not inactivated after 10 min of boiling, but is destroyed by digestion with pronase or trypsin.

The later investigations of Nagasawa *et al.* (1979), involving the use of highly purified PTTH preparations, confirmed its proteinaceous nature, and gel filtration indicates that its molecular weight is about 4400. Since methods of PTTH purification became standardized (see Chapter 3) the results have been reproducible, and this augurs well for further studies on the chemical structure of the hormone.

4.2.1.1. Influence of environmental factors on the production and release of PTTH

As already mentioned, insect development is characterized by cyclic growth and moulting in the successive instars. In each of the instars, different tissues are stimulated to development by cyclically secreted ecdysone, and this will be discussed later. Ecdysone is secreted by the prothoracic glands, and these in turn are activated by PTTH. The question thus arises as to what causes PTTH release.

Recent investigations indicate that the hormone is produced by the lateral cells of the insect brain (Gibbs and Riddiford, 1977; Agui *et al.*, 1979; Riddiford, 1980b; Muszyńska-Pytel, 1987), suggesting the possibility that its activity may be controlled by nerve connections with the visual organs, and the involvement of the optic lobes. However, little is known about the factors controlling PTTH release. In general, it is assumed that control is mainly by environmental factors such as photoperiod, temperature and food.

As shown by the early studies of Wigglesworth (1934), the development and cyclic moulting of *Rhodnius prolixus* feeding on blood are stimulated by the intake of enough blood to cause significant distension of the abdomen. Consumption of a small amount of blood does not stimulate development, suggesting that it is not the chemical composition of the food that is important, but rather stimuli arising from the mechanical stretching of the gut. Cutting the nerve cord at the level of the thorax prevents moulting in spite of the consumption of a suitable amount of food. This proves that electric pulses arising in the mechanoreceptors located in the gut or in other parts of the abdomen reach the brain and evoke rapid release of PTTH. This hypothesis was confirmed by the investigations of Nijhout (1979) using *Oncopeltus fasciatus.* This author demonstrated that development can be stimulated by injecting physiological saline into the abdomen, stretching the body in the same way as a blood meal. This has been confirmed by histological examination of neurosecretory brain cells after starvation followed by *ad libitum* feeding (Morris and Steel, 1977).

The mechanism of PTTH release described here is characteristic not only for *Rhodnius prolixus*, but also for some ticks (Gabbay and Warburg, 1976), and is evidence of the adaptation of these animals to the habit of feeding on blood. Many other data point to the important role of proprioreception in the mechanisms of PTTH production and release. It has been demonstrated that restriction of space or immobilization of the larva leads to a considerable delay in, or complete inhibition of, pupation (Woolever and Pipa, 1970; Giebułtowicz *et al.*, 1980; Boguś and Cymborowski, 1981). Implantation of brains from insects reared normally brings resumption of metamorphosis.

Nijhout and Williams (1974) and Nijhout (1975) consider that the decisive factor causing PTTH release in the last larval stage of *Manduca sexta* is the attainment of a critical body weight. According to these authors, when the critical weight of 5 g (Nijhout, 1975) is reached, a process starts which is not clearly understood, and which requires 24 h for its completion. When this process has ended, the brain is "ready" for PTTH release. This, however, does not occur until a "gate" opens, and this is dependent on photoperiodic conditions (Truman, 1972; Truman and Riddiford, 1974). Figure 4.2 illustrates the interrelations between the photoperiodic conditions and the secretion of PTTH into the haemolymph during development of the last larval instar of *Manduca sexta.* Release of this hormone triggers a number of changes, both in behaviour and morphology.

It seems, however, that the attainment of critical body dimensions does not determine the future fate of the larva. Investigations on the last larval instar of *Galleria mellonella* showed that cold stress at the beginning of this stage leads to the development of additional larval stages, although normally the larvae would pupate (Cymborowski and Boguś, 1976). In the case of *Manduca sexta* the factor inducing additional larval stages is starvation (Cymborowski *et al.*, 1979). In both cases the factor applied produces an extreme increase in the concentration of juvenile hormone in the haemolymph of the insect.

Beside the mentioned effect synchronizing the individual stages of develop-

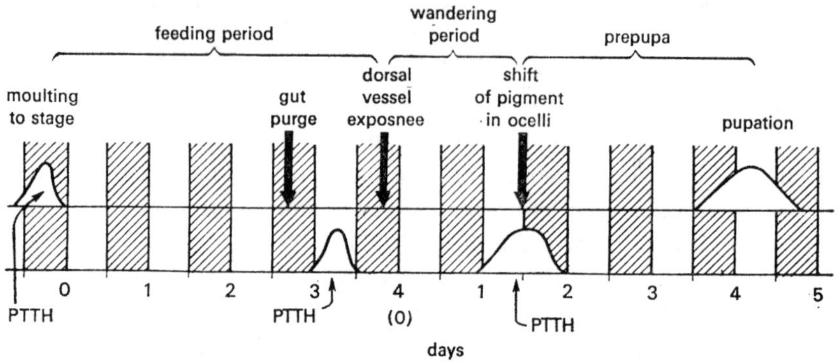

Figure 4.2. Synchronization of morphological changes and behaviour in the last larval instar of *Manduca sexta* (upper part of diagram) and associated periods of PTTH secretion into haemolymph (lower part). Hatched areas—periods of darkness (LD 12:12). First PTTH release starts full chain of endocrine and morphological processes. After Truman and Riddiford (1974), with permission.

ment of the insects, the light conditions also essentially influence the period of larval development (Kryspin *et al.*, 1975; Cymborowski and Giebułtowicz, 1976; Truman, 1976; Cymborowski 1977).

An important factor delaying or accelerating metamorphosis is population density. Corbet (1971) studied the duration of the prepupal stage in *Ephestia kuehniella* reared singly or in groups of increasing density. It was found that both single larvae and those kept in high-density groups show a considerable delay in pupation. Similar results were obtained by Cotter (1974).

Metamorphosis of *Manduca sexta* larvae is also greatly delayed if, after the moult to the last instar and the beginning of the wandering period, they are kept in crowded conditions (Cymborowski, unpublished data). It would seem that in this situation a factor associated with social nature inhibits PTTH release. These larvae, when kept in isolation, begin metamorphosis immediately.

Rearing of crickets (*Acheta domesticus*) in great numbers and in crowded conditions, on the other hand, hastens their development (Johnston and McFarlane, 1973). Similarly, *Dermestes vulpinus*, under conditions of crowding, has a shortened larval period and, more importantly, moults synchronously with the other individuals in the group (Sobczyk *et al.*, 1979). This is indirect evidence that crowding affects the endocrine function of the brain.

Crowding also induces diapause, and thus inhibits metamorphosis. For instance, pupal diapause is caused by periodic inhibition of PTTH secretion by the brain, and its occurrence depends on whether the hormone's action is unblocked; this occurs when the pupae are exposed to low temperature or to a long photoperiod (Williams, 1946). A special chapter of this book will be devoted to the problems of diapause.

Humidity is another environmental factor that influences the development and metamorphosis of some insects. Excessive humidity greatly delays pupation of *Sarcophaga peregrina* larvae (Ohtaki, 1966). This is not due, however, to the direct effect of water on formation of the new cuticle, since the larvae can pupate under humid conditions if they have previously been kept in a dry atmosphere. Transfer of larvae from extremely moist conditions, in which they are unable to pupate, to a dry place leads to pupation within 18–24 h. This is one more indication of the influence of the environment on the secretory function of the brain. This work showed that the delay in pupation observed in the insects was caused by a lack of ecdysone in the haemolymph, secretion of which is stimulated by PTTH.

4.2.1.2. Mechanisms of PTTH release

The data presented here show that environmental conditions play an important role in insect development and metamorphosis. The question therefore arises as to the mechanism of action of these factors on the insect endocrine system. Relatively little research has been done on these problems; nevertheless, an outline can be traced. It has been found, for example, that the insect brain contains extraoptical centres sensitive to light (Tanaka, 1950). Later Lees (1960, 1964) and Steel and Lees (1977), by applying selective illumination or by destroying various parts of the aphid brain, demonstrated that the region of the *pars intercerebralis* in which the neurosecretory cells are located is the part that is sensitive to light. Williams and Adkisson (1964) and Claret (1966) demonstrated that the brain retains its sensitivity to light even if it is transplanted to the abdomen of the insect. It became obvious from these experiments that light conditions may directly modify the secretory activity of the insect brain, but we still do not know the mechanism of this action. Some investigators suggest that the pigments present in the brain may constitute an important link in the sensitivity of the insect brain to light (Zimmerman and Goldsmith, 1971).

In *Spodoptera littoralis* distinct changes are observed in the distribution of pigment in the brain of the final larval instar, distribution of pigment changes during metamorphosis (Plate 18). This suggests the involvement of the pigment in the action of light on brain function. Electrophysiological investigations of the brain of *Periplaneta americana* demonstrated that the medial neurosecretory cells react to changes in illumination by a change in their bioelectric activity (Cook and Milligan, 1972). These studies, together with the results of electrical stimulation of the neurosecretory cells of the brain of *Locusta migratoria* (Girardie and De Reggi, 1978), give grounds for supposing that, under natural conditions, electric pulses originating in various receptors of the insect significantly affect the secretory activity of the brain. Thus, on the one hand they may inhibit or completely block PTTH secretion, but, on the other hand, they may stimulate it. There is in fact much evidence that PTTH secretion may be blocked by the nervous system. This is indicated by the results of studies on the diapause of some insects. For instance, when brains of diapausing *Antheraea polyphemus* and *Hyalophora cecropia* pupae are transplanted to their own abdomens, diapause is interrupted (McDaniel and Berey, 1976). Thus, interruption of the nervous connections of the brain activates its secretory activity, indicating the inhibitory influence of these connections. The brain of diapausing *Mamestra brassicae* pupae also activates the prothoracic glands under *in vitro* conditions (Agui, 1975).

Recently attention has been also drawn to the possibility that biogenic amines are involved in the accumulation and secretion of PTTH in the insect brain. Injection of noradrenaline into the haemolymph of *Galleria mellonella* larvae results in accumulation of neurosecretory material in the type *A* neurosecretory cells of the *pars intercerebralis* of the brain (Wartoń and Dutkowski, 1977). At the same time secretory material of aminergic origin disappears from these cells (Wartoń and Dutkowski, 1978). Muszyńska-Pytel and Cymborowski (1978) suggest that serotonin may release neurosecretory material from cricket brains. Scharrer and Wurzelmann (1978) similarly observed enhanced release of neurosecretory granules in the *corpora cardiaca* of *Leucophaea maderae* as early as 3 min after injection of serotonin into the haemolymph. It therefore seems likely that in the near future the role of biogenic amines in regulation of PTTH release will be known in greater detail.

4.2.1.3. Mechanisms of action of PTTH

Although the complete purification of PTTH has not yet been achieved, much information has accumulated on its mechanism of action. In the first place much indirect and direct evidence is available that the prothoracic glands are the target tissue for the action of PTTH (Herman, 1967). It has been shown, for instance, that this hormone stimulates RNA synthesis in the prothoracic glands of the silkworm and cockroach (Oberlander *et al.*, 1965; Gersch and Stürzebecher, 1970). Under the influence of this hormone a lipid fraction, the structure of which is not known in detail, appears in the heamolymph (Gersch and Stürzebecher, 1970).

The first step in the activation of the prothoracic glands seems to be an increase in cyclic AMP (cAMP) within the cells of the glands. It has been demonstrated by Vedeckis and Gilbert (1973) that the prothoracic glands of *Manduca sexta* produce ecdysone *in vitro* under the influence of cAMP even in the presence of inhibitors of phosphodiesterase, the enzyme inactivating cAMP. These authors also demonstrated that in the prothoracic glands the cAMP concentration reaches its maximum during the period when PTTH is released in order to initiate metamorphosis. It seems, therefore, that PTTH acts in a similar way to the hormones of the anterior pituitary in vertbrates, stimulating the synthesis and release of steroid hormones.

4.2.2. Ecdysone

As already described, information from the external environment is transformed in the neurosecretory cells of the insect brain to chemical information in the form of the neurohormone PTTH. Secretion of this hormone in the appropriate amount and at the proper time into the heamolymph decides the subsequent fate of the insect. PTTH stimulates the prothoracic glands to produce and release ecdysone, the hormone of metamorphosis.

4.2.2.1. Forms of ecdysone

In 1954 Butenandt and Karlson developed a method for the extraction of ecdysone from silkworm (*Bombyx mori*) pupae, and during the course of the next decade the chemical structure of the hormone was determined (Karlson *et al.*, 1965; Hüber and Hoppe, 1966). From one ton of silkworm pupae 250 mg of ecdysone was obtained. Further purification yielded 25 mg of the crystalline form. It appeared that ecdysone and its more active form, 20-hydroxyecdysone, are derivatives of cholesterol (Fig. 4.3). At present 20-hydroxyecdysterone is isolated from various plants, in which it is present in large amounts. For instance, 1 g dry weight of yew (*Taxus*) needles contains 500 μg of pure 20-hydroxyecdysone; 70 g of dry needles therefore contain as much biologically active substance as 1 ton of silkworm pupae. The occurrence of ecdysone in various sources has caused some confusion in terminology: β-ecdysone, polipodin A, ecdysterone

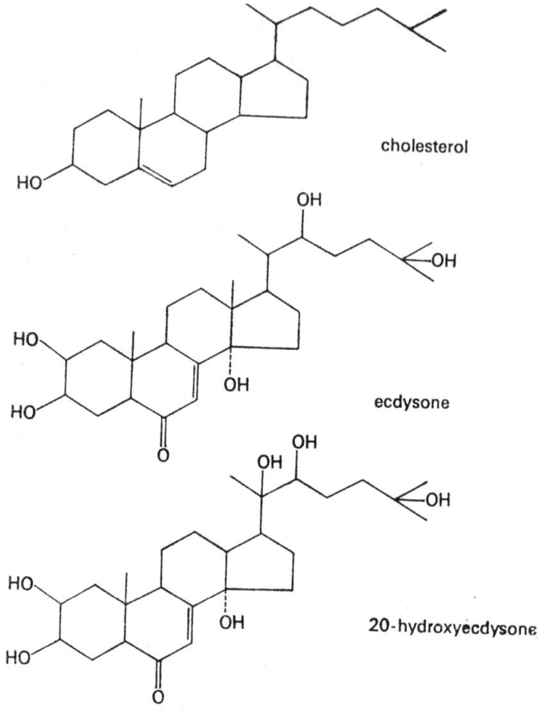

Figure 4.3. Chemical structures of ecdysone and 20-hydroxyecdysone compared with structure of cholesterol, from which these hormones arise.

and crustacesterone are all synonyms for the same chemical compound—20-hydroxyecdysone (Horn *et al.*, 1966). For this reason the term ecdysteroid is used in this book in the general sense, to include the various kinds of ecdysone.

Nakanishi *et al.* (1966) isolated from the plant *Podolarpus nakii* an ecdysone-like substance in surprisingly large amounts, and called it phytoecdysone. At present about 50 phytoecdysones are known, among them 20-hydroxyecdysone, which also occurs in insects. The concentration of these hormones may be as high as 1 per cent of the dry weight of the plant, and hence it is plants rather than insects that are the source of the ecdysteroids used in endocrinological studies. Why do plants produce ecdysteroids? There is no good answer. It is supposed that their presence in plants may be a defence against insect attack, and in many cases this may be true. However, insects have a very efficient mechanism for the degradation and removal of ecdysteroids from the body. The silkworm (*Bombyx mori*), for instance, thrives on plant species containing large amounts of ecdysteroids (Hikino and Takemoto, 1974).

Intensive research over many years led to the establishment of the chemical structures of six ecdysteroids (Fig. 4.4). As many as five of them were identified in *Manduca sexta*. Further types were isolated from crustaceans. For many years there was no agreement as to whether ecdysone or 20-hydroxyecdysone is produced by the prothoracic glands. The problem now seems to be resolved, as a result of *in vitro* investigations. It proved that *in vitro* the PG produce only ecdysone (King *et al.*, 1974), which is further transformed (hydroxylated) to 20-hydroxyecdysone by tissues such as the fat body, Malpighian tubules, gut and epidermis (Mayer *et al.*, 1978).

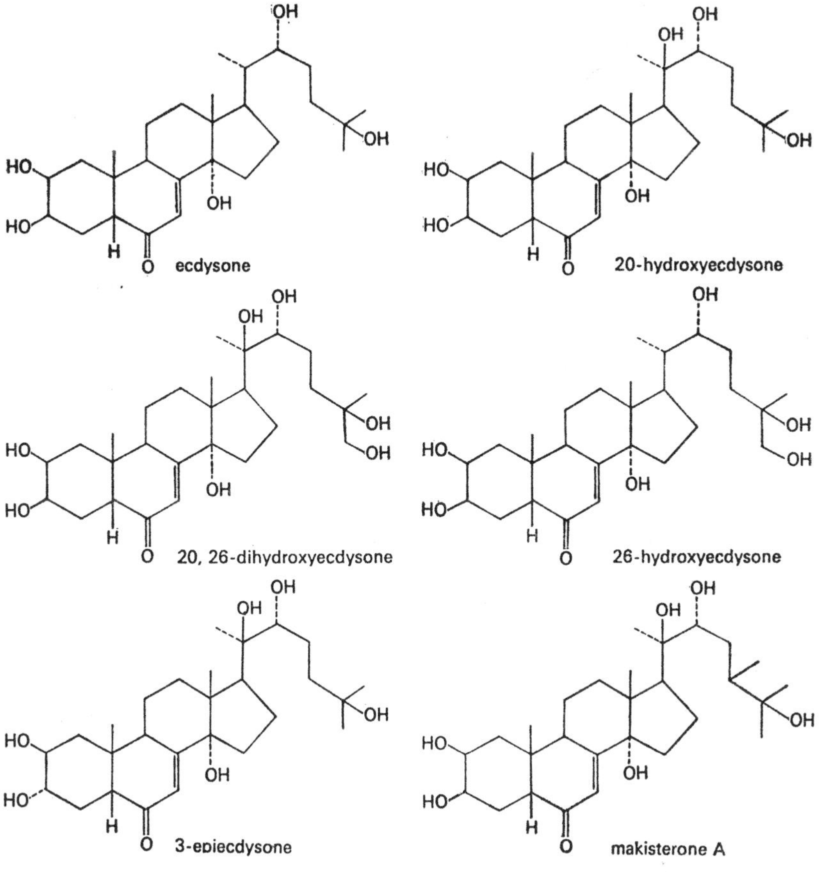

Figure 4.4. Forms of ecdysteroids isolated from various insects.

In vitro studies also showed that 20-hydroxyecdysone stimulates morphological changes at a concentration as low as 10^{-7} M/ml, whereas a 100-fold greater concentration of ecdysone is necessary to produce similar changes. Both these forms occur in insects at the very low concentrations of 10^{-5}–10^{-9}% of fresh insect weight.

There has recently been an increase in the number of reports on ecdysone production by various insect tissues. Gersch and Eibisch (1977), for instance, found that ecdysone is synthesized from cholesterol in *Periplaneta americana* even when the prothoracic glands are removed. The ovaries and eggs of the wax moth (*Galleria mellonella*) contain as many as five different ecdysteroids (Hsiao and Hsiao, 1979). The ovaries of the mosquito *Aedes aegypti* produce ecdysone, and the process is controlled by the brain (Hagedorn *et al.*, 1979).

4.2.2.2. Changes in ecdysteroid level during insect development

What is the function of ecdysone in the insect body? To answer this question the level of this hormone in the haemolymph of insects in various developmental stages should be examined. The ecdysone concentration depends on several main factors: the rate of production of the hormone, the rate at which it is transformed into 20-hydroxyecdysone, the rate of uptake by the tissues, and the rate of degradation. The first measurements of the concentration of ecdysone in the haemolymph of lepidopteran insects were performed by Shaaya and Karlson (1965). These authors demonstrated the concentration peaks before each larval moult, whereas the hormone is completely absent during the remainder of the period between moults. At the time of Shaaya and Karlson's work, ecdysone and 20-hydroxyecdysone had not yet been distinguished. Today, highly sensitive radioimmunological assay methods are available for both these hormones (see Chapter 3), allowing precise monitoring of changes in their relative proportions during development.

Changes in ecdysteroid levels in the body of hemimetabolous and holometabolous insects are shown in Figure 4.5. In the locust (Fig. 4.5 A) ecdysone appears two days before the last larval moult and three days before imaginal moult. On the other hand, in *Holometabola* such as *Manduca sexta* (Fig. 4.5 B) two ecdysteroid peaks appear towards the end of the last larval stage (Lafont *et al.*, 1974; Bollenbacher *et al.*, 1975). During the first rise in concentration the hormone consists of both ecdysone and 20-hydroxyecdysone, in a ratio of 1:1. The function of the hormone is to evoke changes in the behaviour of the larva. After release of ecdysone the larva ceases to feed and searches for a place suitable for pupation.

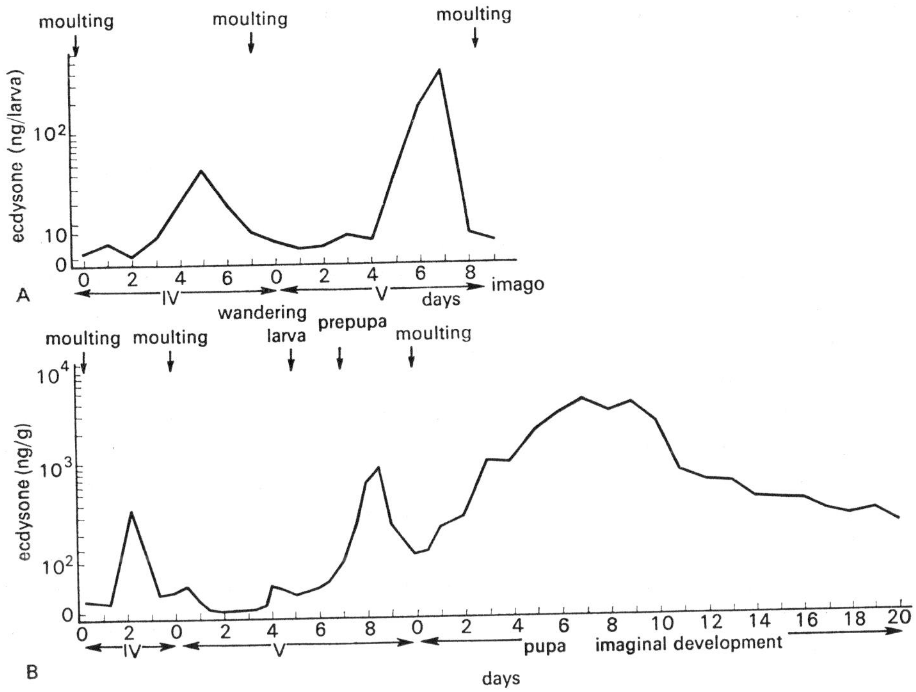

Figure 4.5. Changes in ecdysone concentration in the body during development and metamorphosis (A) incomplete, (B) complete. After Riddiford and Truman (1978), with permission.

Another very important function of ecdysone during this minor peak is to initiate the change from the larval to the pupal "programme" in the cells of the epidermis (Riddiford, 1978). Under normal conditions, ecdysone causes the epidermis to produce only pupal cuticle.

The second ecdysteroid peak is much higher than the first, and the ratio of ecdysone to 20-hydroxyecdysone is 1:5. This second appearance of the hormones in the haemolymph brings about the degeneration of the old larval cuticle and the formation of a new pupal cuticle. Under the influence of ecdysteroids, other larval organs are also transformed into pupal structures. After pupation there is a third ecdysteroid peak, which initiates imaginal development. Usually, a high concentration of the hormones persists for about one-third of the duration of imaginal development. Experiments involving the use of ligatures indicate that the presence of ecdysteroid is indispensable for the normal completion of metamorphosis. Since the prothoracic glands atrophy during the pupal stage, it would seem that in this period ecdysone is produced by other tissues (see p. 86).

4.2.2.3. Inactivation of ecdysone

As seen in Figure 4.5, ecdysone appears very rapidly in the body of the insect and disappears in the same way. This indicates that insects must have an efficient means of inactivating this hormone. Shaaya (1969) reports that the half-life of 20-hydroxyecdysone in *Calliphora* is 3 h while Ohtaki *et al.* (1968) noted that when 1 μg ecdysone is injected into mature *Sarcophaga peregrina* larvae, 50 per cent of the injected hormone disappears within one hour, and 98 per cent has disappeared after 8 h.

One of the methods by which insects rid themselves of ecdysone is to excrete it unchanged. Most investigators, however, affirm that it is oxidized, principally by the fat body, leading to biologically inactive compounds (Karlson and Bode, 1969). To date there is no agreement as to the nature of these compounds. Injection of labelled ecdysone into the haemolymph of the locust (*Locusta migratoria*) leads to excretion of 3-dehydroecdysone, 20-hydroxyecdysone and 3-dehydro-20-hydroxyecdysone (Koolman *et al.*, 1975). Intensive studies are now under way with the aim of establishing the metabolic fate of ecdysone in various insect groups.

4.2.2.4. Mechanism of action of ecdysone

Clever and Karlson (1960) demonstrated that ecdysone acts directly on the genome of insect cells. They observed that ecdysone causes the appearance of puffs at specific loci of polytene chromosomes in the salivary glands of *Chironomus tentans* (Fig. 4.6). In one such region, band J18C, a puff arises as early as 15 min after the administration of ecdysone. A few minutes later puff IV 2B arises. Further puffs appear much later, after about 5 h of hormone action. It was noted that, during normal development of the insect, all these puffs appear during the pupal stage, when the ecdysone level is very high. If actinomycin D, an inhibitor of RNA synthesis, is administered before treatment with ecdysone, puffs do not appear. It is interesting that when protein synthesis inhibitors are administered, the appearance of the first puffs, which occurs within 15 min of ecdysone treatment, cannot be arrested. These puffs, however, inhibit later ones. It would seem, therefore, that ecdysone only activates the formation of the first puffs, which produce a substance activating further DNA segments.

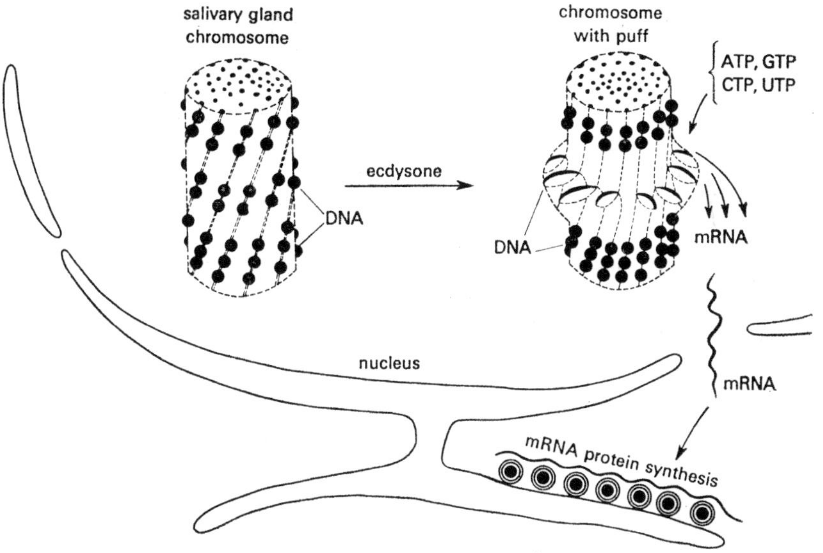

Figure 4.6. Mechanism of action of ecdysone according to P. Karlson. The hormone acts directly on DNA, forming a puff. Here is synthesized mRNA, which then migrates to the cytoplasm, where specific protein is produced as the result of the action of ecdysone on the target cell. After P. Karlson. Perspec. Biol. Med. **6**; 203–214 (1963).

Later, an alternative hypothesis on the mechanism of action of ecdysone was advanced (Karlson and Koolman, 1976), which assumed the existence of cytoplasmic and nuclear receptors in the target cell (Fig. 4.7). In 1978 two papers appeared simultaneously in the same journal reporting the detection of ecdysteroid receptors. Detection of these receptors became possible when the tritiated ecdysone analogue ponasterone A became available; this compound is at least 100 times more active than 20-hydroxyecdysone. More significantly, it was possible to prepare radioactive ponasterone A with a specific activity of 126 Ci/mM (Cymborowski, 1982).

Using this analogue Maroy *et al.* (1978) conducted *in vitro* investigations on *Drosophila melanogaster* cells (Kc cells). Incubation of these cells in the presence of ^{3}H-ponasterone A leads to specific binding of the analogue both to the cytosol fraction and to the nuclear fraction. It would seem, therefore, that ecdysone first enters the cell, where it binds with the cytoplasmic protein receptor, and is then transported to the nucleus in the form of a receptor-hormone complex.

On account of their very small number (80–500), the cytoplasmic and nuclear receptors could not be characterized in detail. The affinity coefficient of ponaste-

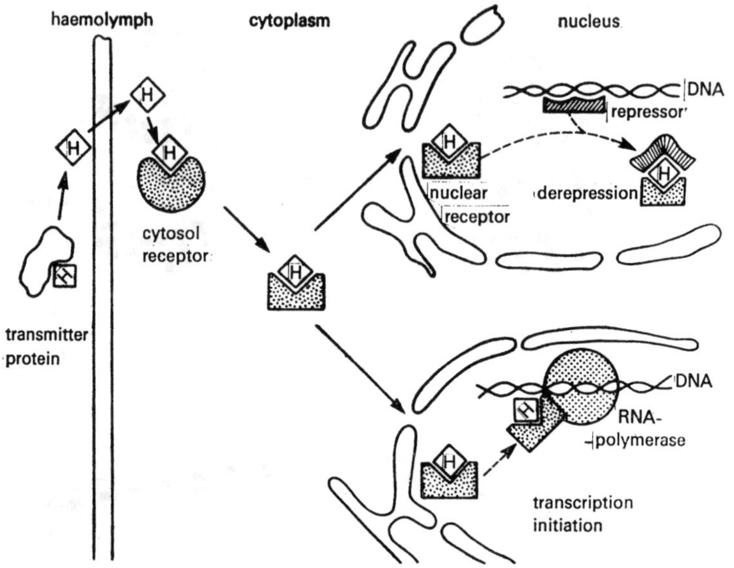

Figure 4.7. Alternative model of the action of ecdysone on target cells, taking into account protein receptors. Details in text. After Karlson and Koolman (1976), with permission.

rone A was found to be 3×10^{-9} M, both for the cytosol fraction and for the nuclear fraction. For the nuclear fraction the affinity coefficient for ponasterone A was 7×10^{-10} M.

In other studies (Yound *et al.*, 1978; Yound, 1980) ponasterone A was used to investigate ecdysteroid receptors in imaginal discs of *Drosophila melanogaster*. In these investigations 10 000–20 000 discs/ml incubation fluid were used, with an incubation time of 90 min. In contrast to the study described in the previous paragraph, no difference was found here in the binding of ponasterone A by the nuclear and cytosol fractions. In both cases the affinity coefficient was $3–4 \times \times 10^{-9}$ M for ponasterone A, and was in agreement with data on the morphogenetic action of this hormone. It was found that the lowest ponasterone A concentration producing morphological changes in the discs *in vitro* is 4.2×10^{-9} mol/ml.

As in the previously described study, the very small number of receptors detected made it impossible to characterize them in any detail. The molecular weight of the protein was found to vary between 250 000 and 480 000. In these investigations no differences in the effectiveness of binding were found between the cytoplasm and the cell nucleus, suggesting that the cytoplasmic and nuclear receptors in the discs are identical.

Kc cells are being studied intensively as regards the mechanism of action of ecdysone. Cherbas *et al.* (1980a, b) consider that these cells are very convenient for this purpose, and new data from this field can no doubt be expected in the near future.

4.2.3. Juvenile hormone

Wigglesworth (1934) was the first to present evidence of the existence in the head of insects of a factor preventing metamorphosis. When the haemolymph of a young last-instar *Rhodnius prolixus* larva was connected to that of an older one preparing for metamorphosis, the metamorphosis of the older insect was greatly delayed. In some cases the older larva underwent an additional larval moult, or an intermediate form between the larva and the imago was formed. In further investigations the same author demonstrated that the inhibitory factor is produced by the *corpora allata* (*CA*).

After these discoveries numerous other communications appeared, such as those by Bouhniol (1936) and Piepho (1940), who obtained similar results on species of the *Lepidoptera*, i.e. insects undergoing full metamorphosis. It further appeared that removal of the *CA* in early larval stages leads to premature metamorphosis—a miniature pupa is formed—whereas implantation of *CA* into last-instar larvae inhibits metamorphosis leading to the formation of very large superlarvae.

In 1956 Williams obtained from the abdomens of adult males of *Hyalophora cecropia* a substance he described as "a golden oil". It markedly inhibited metamorphosis, and when given to freshly formed silkworm pupae caused formation of further pupae instead of imagines.

In subsequent years much work was done to establish the chemical structure of the substance detected by Williams. The structure of this compound, called the juvenile hormone (JH), was determined by Röller *et al.* (1967). Soon afterwards it was found that as many as four forms of the hormone occur in insects (Fig. 4.8). The chemical structure of JH O was presented as recently as 1980 at the international conference in Karpacz (Bergot *et al.*, 1981). This form was detected in *Manduca sexta* embryos. The remaining three forms, JH I, JH II, JH III, are present in various proportions in individual developmental stages of insects. It is not yet clear what is the function of each of these forms. It was, however, found that, in the haemolymph of *Manduca sexta* larvae, three forms of the hormones are present (JH O being absent), JH I and JH II being the prevalent forms, whereas in adult females JH II and JH III usually occur at a ratio of 1:1

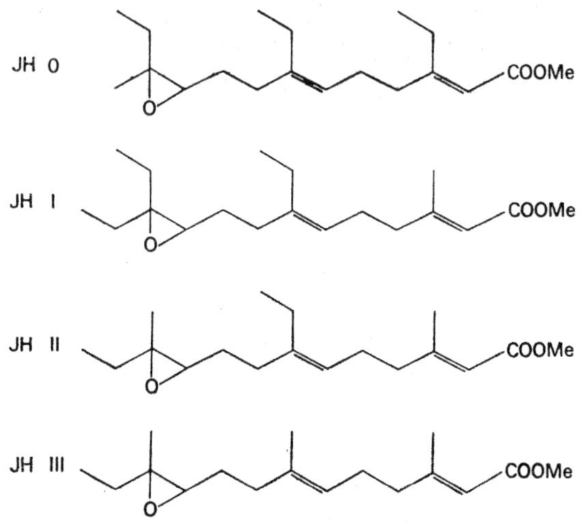

Figure 4.8. Four types of juvenile hormone isolated from various insects.

(Judy *et al.*, 1973; Reibstein and Law, 1973; Fain and Riddiford, 1975). Thus, it seems that the main functions of JH III are gonadotropic (see Chapter 5).

4.2.3.1. JH concentration in the insect body

As for ecdysone, the level of JH in the haemolymph determines its biological activity. Numerous highly sensitive assay methods have recently been developed for JH (see Chapter 3). Juvenile hormone concentration depends, of course, on the activity of the *corpora allata*, and also on that of degrading enzymes, particularly the specific JH esterases.

Juvenile hormone concentration is high during the periods of larval moulting; it then decreases during the time that the larvae are feeding. In the last larval instar of *Manduca sexta* the JH concentration is relatively high for more or less the first half of the feeding period (Fig. 4.9). It then falls drastically to an undetectable level (Fain and Riddiford, 1975). This decline in the concentration of JH in the haemolymph is caused by a simultaneous decrease in the secretory activity of the *CA* and the appearance of JH esterases, both in the haemolymph and in various tissues (Riddiford, 1980a). As we shall see later, the disappearance of JH from the haemolymph is decisive for the change from the larval to the pupal programme. This change requires that the mechanisms leading to pupation are "switched on".

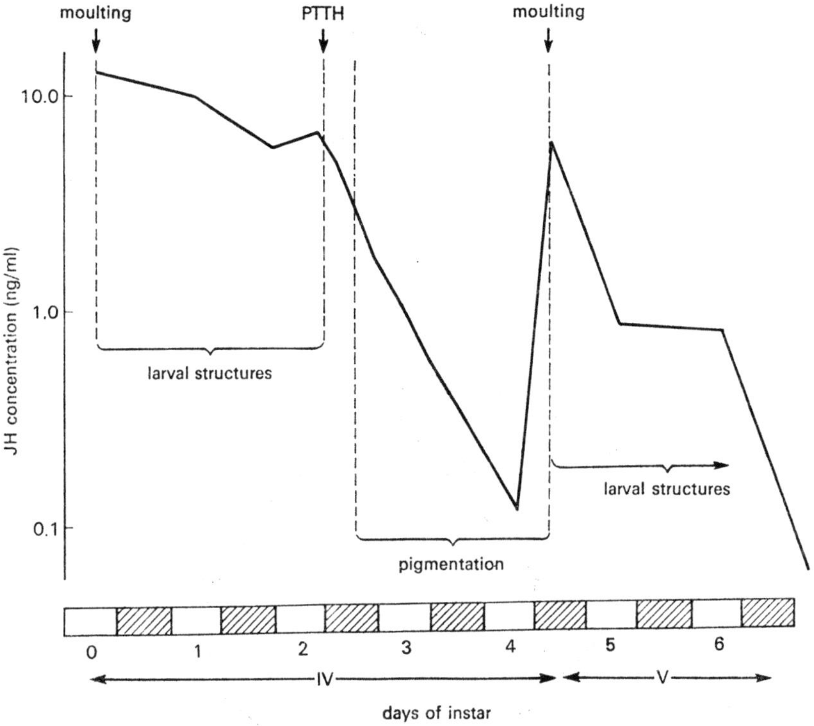

Figure 4.9. Changes in juvenile hormone level in haemolymph of final instar larvae of *Manduca sexta*. PTTH—prothoracicotropic hormone, IV—penultimate stage, V—last stage of larval development. Modified after Fain and Riddiford (1975), with permission.

4.2.3.2. Mechanisms of regulation of JH level

The *corpora allata* are switched off towards the end of larval life, except in species that have a larval diapause. They resume their activity shortly before pupation (Cymborowski and Stolarz, 1979).

The next period of *CA* activity occurs during imaginal life. In females the *CA* are cyclically switched on and off depending on the developmental stage of the ovaries (see Chapter 5).

Unfortunately, the mechanism controlling *CA* activity is not yet known, though it is generally considered that the brain plays the most important role. For instance Granger and Sehnal (1974) and Sehnal and Granger (1975) claim that the *CA* are activated in *Galleria mellonella* by a neurohormonal factor originating in the larval brain, whereas their inactivation before the beginning of metamorphosis occurs through a nervous pathway.

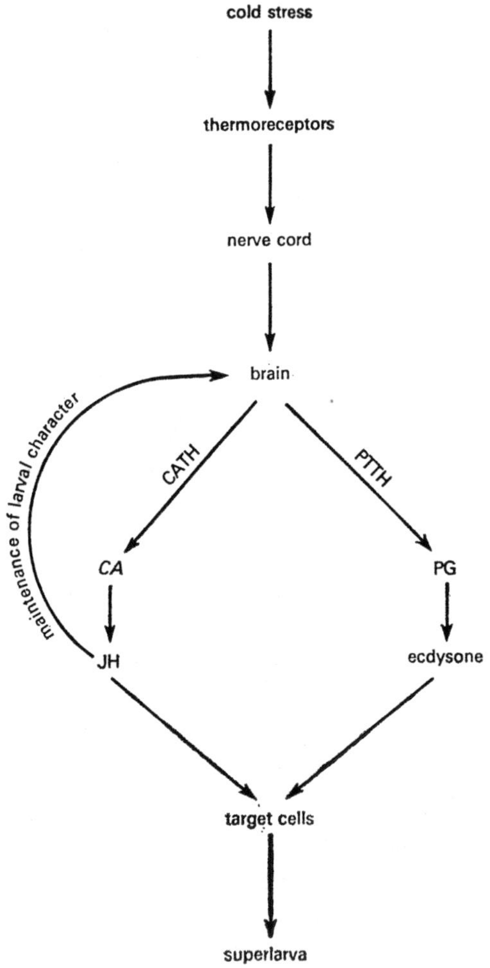

Figure 4.10. Diagram showing effect of cold stress on functions of endocrine system of last instar *Galleria mellonella* larvae. Information on thermal state of body reaches brain via thermoreceptors and causes delay in switching of "larval development programme" to "pupal development". This leads to release by brain of an allatotropic factor (CATH) which stimulates *corpora allata* (*CA*) to produce JH. As a result of chilling the brain preserves its larval character, releasing PTTH in the presence of JH; this would be impossible under normal conditions of development of the last larval instar. PTTH release causes development of superlarvae. After Boguś and Cymborowski (1981).

Summing up the results of studies on changes in *CA* activity in *Manduca sexta* larvae, Williams (1976) advances the following hypothesis for the regulation of the activity of these glands. When the last-instar larva reaches a weight of about 5 g, a substance referred to as an inhibitor is released from an unknown source in the abdomen, and acts upon the brain. The brain in turn switches off the *CA* via a nervous pathway. Switching on of the *CA* results from the production by the brain of a stimulating substance. Thus, activation of the *CA* would depend on the appearance of an unknown allatotropic factor, and depression of *CA* activity would result from the release of the inhibitory factor (allatohibin). The author describes the *CA* inhibitor isolated from *Manduca sexta* larvae as a thermolabile substance sensitive to trypsin, with a molecular weight around 1400. To date, however, no investigations have been undertaken to elucidate the mechanism of action of this unknown substance. From the theoretical and perhaps also the practical point of view it would be extremely important to obtain information on this mechanism.

Galleria mellonella seems to be an excellent model for the investigation of the allatotropic factor since, as already mentioned, periodic chilling of final-instar

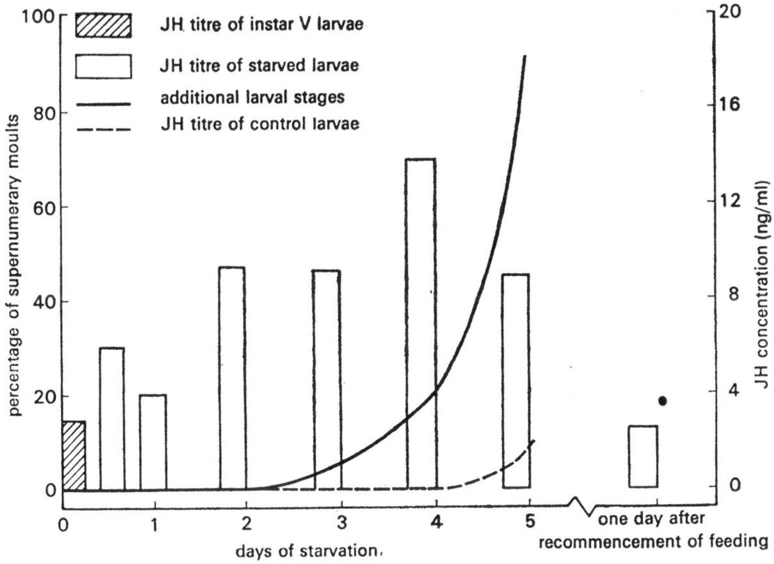

Figure 4.11. Effect of starvation of last instar *Manduca sexta* larvae on development of supernumerary larval stages (continuous line) and on JH level in their haemolymph (open columns) and changes of JH level in normal last instar larvae (broken line). Note sharp fall in JH level after recommencing feeding of the larvae. After Cymborowski *et al.* (1979).

larvae evokes release into the haemolymph of large quantities of JH, leading to the appearance of additional larval stages (Boguś and Cymborowski, 1981). If stimulation of the *CA* occurs by the hormonal route, larvae should therefore contain most hormone after exposure to cold (Fig. 4.10).

A similar mechanism seems to operate during starvation of last-instar *Manduca sexta* larvae. Starvation induces a rise in the level of JH in the haemolymph of these insects (Fig. 4.11), whereas feeding immediately depresses the hormone level. This appears to be a good model for investigation of the mechanism of *CA* inactivation.

4.2.3.3. Mechanism of action of JH

Numerous investigations have shown that carrier proteins specifically binding JH are present in insect haemolymph. For instance Ferkovich *et al.* (1975) demonstrated that in *Plodia interpunctella* larvae there is a protein with molecular weight 2.5×10^4 that selectively binds JH. Similar experiments were performed with *Manduca sexta* (Goodman and Gilbert, 1978). The main function of these proteins seems to be the protection of JH from the action of non-specific esterases occurring in the haemolymph (Rudnicka *et al.*, 1979).

Another, no less important, function of carrier proteins is probably the facilitation of the binding of JH to receptors in target tissues and cells. Since the main site of action of juvenile hormone is the cuticle, it is principally there that the receptors were sought. Schmialek (1973) isolated a protein binding specifically to one of the juvenile hormone analogues from the epidermis of the pupal cuticle of the mealworm (*Tenebrio molitor*). In his experiments, he used tritiated hormone with a specific activity of 28 Ci/mM. This hormone, when introduced into the haemolymph of pupae, was actively accumulated by the epidermal cells of the cuticle (Fig. 4.12). Studies of the physicochemical properties of the receptor indicated that it is a ribonucleoprotein with a molecular weight of 360 000 and an isoelectric point of 4.4. The complex has a high affinity coefficient with respect to the JH analogue used in the experiments (4.4×10^{10} M/ml). The very low concentration of this receptor (1.7×10^{-10} M/ml) makes further characterization impossible. The complex is sensitive to ribonuclease and trypsin.

In the same species, juvenile hormone receptors were found in the ovaries (Schmialek *et al.*, 1975). In these organs, the protein isolated had a molecular weight of 8×10^5 and an isoelectric point of 5.4. Under the experimental conditions used it was thermostable and was precipitated by the action of 5 per cent trichloroacetic acid, and was hydrolyzed by 1 N HCl. Binding to JH was specific,

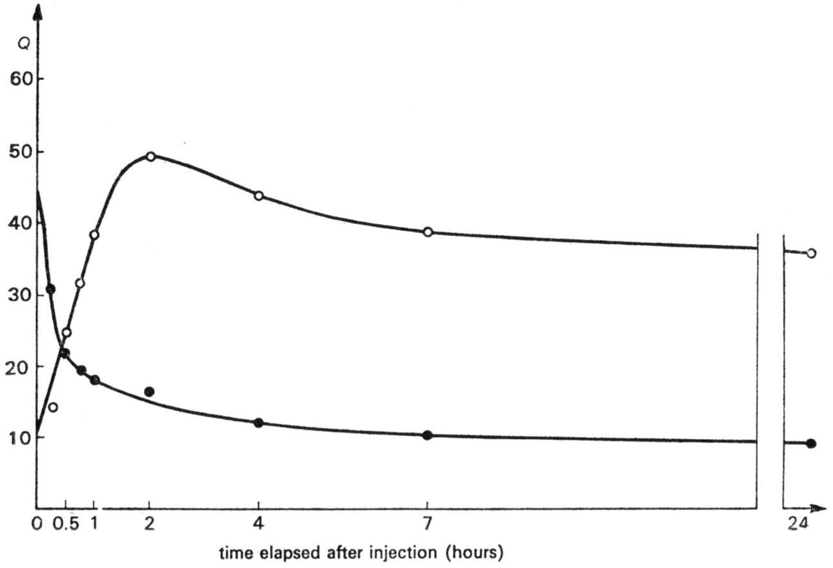

Figure 4.12. Changes in juvenile hormone analogue distribution in epidermis (○) and other parts of the body (●) of *Tenebrio molitor* pupae. The pupae received a single injection of 1 μl of a solution containing 10^{-8} mol/dcm^3 labelled hormone, Q—ratio of non-degraded hormone to total radioactivity as a percentage. After Schmialek (1973), with permission.

and the complex could not be broken by the action of various substances of the same polarity.

This is about all that is known so far about JH receptors. Their location in the cell has not been precisely determined, so that it is not clear whether they occur mainly in the cytoplasm or in the cell nucleus. At present, it is supposed merely that JH acts directly as a co-repressor on the cell genome (Williams and Kafatos, 1971). These authors assume the existence of three main regulator genes which control the functioning of the larval, pupal and imaginal gene sets. In the larval period JH acts as co-repressor of both pupal and imaginal genes. When the JH concentration in the haemolymph falls, the repressor of the regulator gene of the pupa becomes detached, because binding requires a high concentration of JH. When JH disappears completely from the haemolymph, the imaginal genes are unblocked.

Sufficient experimental data either to support or contradict the above hypothesis have not yet been obtained. The unpublished results of Riddiford (personal communication) indicate that JH penetrates the cell and subsequently the cell nucleus, since early in the last larval instar epidermal cells of *Manduca sexta* have a high affinity for tritiated JH, there is still a high concentration of the hormone (in the haemolymph of these larvae), whereas the epidermal cells of older larvae do not show any affinity. As we already know, JH is absent from the haemolymph late in the last larval instar. During this period the larvae are not sensitive to exogenous JH. Sensitivity returns at the pupal stage, suggesting that new JH receptors are being then synthesized.

We thus have an additional important conclusion: the effect of JH in the insect depends not only on the concentration of the hormone in the haemolymph and tissues, but also on the sensitivity of the tissues. This will be discussed in Chaper 8.

4.2.4. The eclosion hormone and bursicon

The last steps of the imaginal moult are regulated by the eclosion hormone and bursicon. Eclosion hormone controls the shedding off the old cuticle (a process which is not as simple as it seems), whereas bursicon controls the tanning and hardening of the new cuticle.

4.2.4.1. Mechanism of action of eclosion hormone

The eclosion hormone is produced by specific brain cells towards the end of the insect's imaginal development (Truman, 1978) and accumulates in the *corpora cardiaca*. Release of this hormone takes place on the last day of imaginal development, at a signal from the biological clock set by the photoperiod (Reynolds *et al.*, 1979). Reynolds and Truman (1979) extracted this hormone from the *CC* of *Manduca sexta* and partially purified it. It proved to be a peptide with a molecular weight of 9000.

When injected into pupae towards the end of imaginal development, purified eclosion hormone evokes the series of movements of the abdomen that are typical of those associated with imaginal moulting (eclosion). As demonstrated by Truman (1978), the hormone acts directly on the nervous system, which becomes sensitive to it only on the last day of imaginal development, shortly before moulting. In *Manduca sexta* sensitivity to this hormone suddenly appears about 4 h before release of the eclosion hormone (Reynolds *et al.*, 1979). This may be connected with the genetically coded synthesis of receptors for the hormone. Such receptors, however, have not yet been isolated.

It is extremely interesting that once it has been acted upon by the eclosion hormone, the nervous system completely loses the ability to respond to the hormone (Truman, 1978). It is still more interesting that this occurs suddenly, suggesting that the cells immediately lose the relevant receptors, and that even whole cells may degenerate (Taylor and Truman, 1974).

What is the mechanism of action of eclosion hormone on insect neurons? The investigations of Truman *et al.*, (1976) demonstrated that crude extracts of this hormone induce an increase in cAMP content in the nervous system of the silkworm, whereas purified extracts produce an increase in the cGMP level and have no influence on cAMP. It was also found that injection of cGMP into a pupa has a 100-fold stronger effect than cAMP in stimulating moulting behaviour. It therefore seems that cGMP is the mediator of the action of eclosion hormone. The detailed mechanism of this action is not yet known.

The eclosion hormone also acts on other tissues in the developing insect (Riddiford and Truman, 1978). The appearance of the hormone in the pupa stimulates degeneration of the intersegmental muscles of the abdomen (Fig. 4.13). These muscles are required for the movements necessary in moulting, and have no other function (Lockshin and Beaulaton, 1974). Eclosion hormone also acts

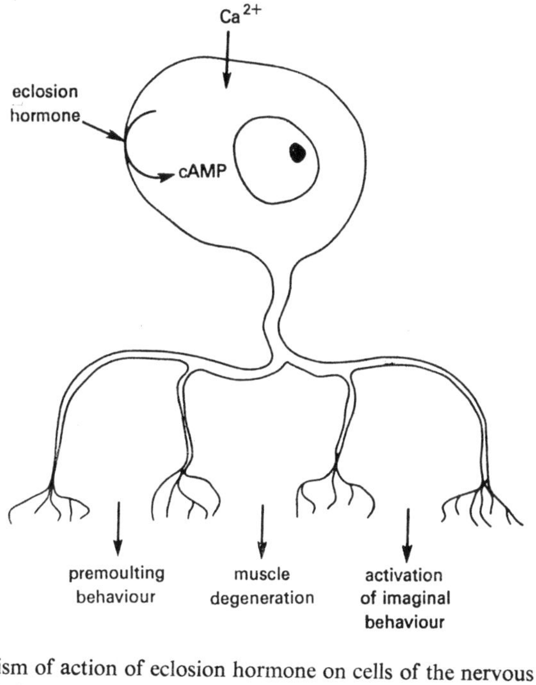

Figure 4.13. Mechanism of action of eclosion hormone on cells of the nervous system. Modified after Riddiford and Truman (1978), with permission.

on the wings of the newly eclosed insect, enhancing their extensibility by softening the cuticle (Reynolds, 1977). In this case, too, eclosion hormone seems to act via cGMP.

4.2.4.2. Mechanism of action of bursicon

Numerous studies have shown that the tanning and hardening of the new cuticle is controlled in most insects by a hormone produced by neurosecretory cells—bursicon (Vincent, 1972; Seligman, 1980). In *Manduca sexta* bursicon is released from the neurosecretory cells of the abdominal ganglia via the perisympathetic system (Truman, 1973). The hormone appears in the haemolymph of the imago immediately after eclosion from the pupa (Reynolds *et al.*, 1979). At this time it acts principally on the wings to maintain their plasticity and allow them to expand completely. Later the wings and whole body become tanned and the cuticle hardens. Bursicon from *Manduca sexta* has been almost completely purified. It is a peptide with a molecular weight of about 40 000. The mechanism of its action is not yet fully known. From studies on *Periplaneta americana* and *Pieris brassicae* it seems that bursicon increases the permeability of haemocytes and epidermal cells to tyrosine (Fig. 4.14). Under the influence of this hormone the permeability of epidermal cells to dopamine also changes. Dopamine is then transformed in the epidermal cells to N-acetyldopamine, and it is this that changes the colour of the new cuticle.

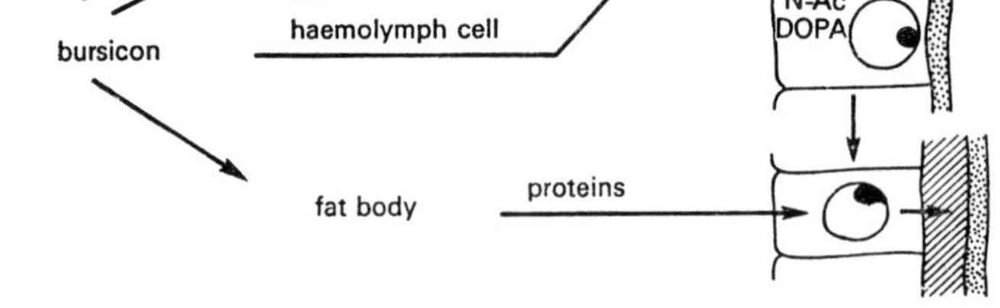

Figure 4.14. Probable mechanism of action of bursicon in the tanning and hardening of the cuticle. DOPA—dopamine, N-AcDOPA—N-acetyl-dopamine. Modified after Riddiford and Truman (1978).

The first step in the action of bursicon is an increase in the concentration of cAMP both in haemocytes and in epidermal cells (Delachambre *et al.*, 1979). However, the mechanism of this process is not yet known.

Bursicon also stimulates deposition of the endocuticle immediately after the imaginal moult. It is believed that this effect involves the synthesis of cuticle proteins by the fat body, or the uptake of these proteins by epidermal cells (Riddiford and Truman, 1978).

4.3. Interaction of hormones during insect development and metamorphosis

From what has been said above, it is evident that ecdysone and juvenile hormone control the growth and metamorphosis of insects, and that PTTH plays a role in the synchronization of these processes. The PTTH release into the haemolymph initiates a number of physiological changes in the insect body. The interaction of these two principal hormones during development and metamorphosis has been demonstrated in *Manduca sexta* (Gilbert *et al.*, 1980; Riddiford, 1980a).

As already mentioned, the ecdysone level rises sharply in this species about two days before each of the larval moults and before pupation (Fig. 4.15). Transition from the pupa to the adult form requires a much higher concentration of ecdysone in the haemolymph, and this is usually observed in the middle of the

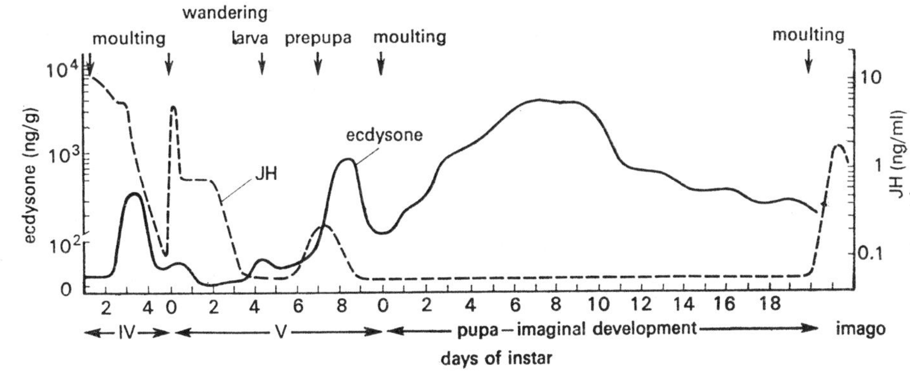

Figure 4.15. Changes in JH and ecdysone levels during larval development and metamorphosis in *Manduca sexta*. After Riddiford (1980), with permission.

period of imago formation. A slight increase in the concentration of ecdysone on the third day of the last larval instar causes a change in the behaviour of the insect. Some authors consider that this is due to the complete disappearance of juvenile hormone at this time. As seen in Figure 4.15, this situation occurs for the first time in the insect's life in the last instar of larval development, and this is one of the ways in which the last instar differs from the previous ones. After the first moderate surge of ecdysteroids, there is another, much higher surge. The amount of hormone released is sufficient to transform the larval structures into pupal ones.

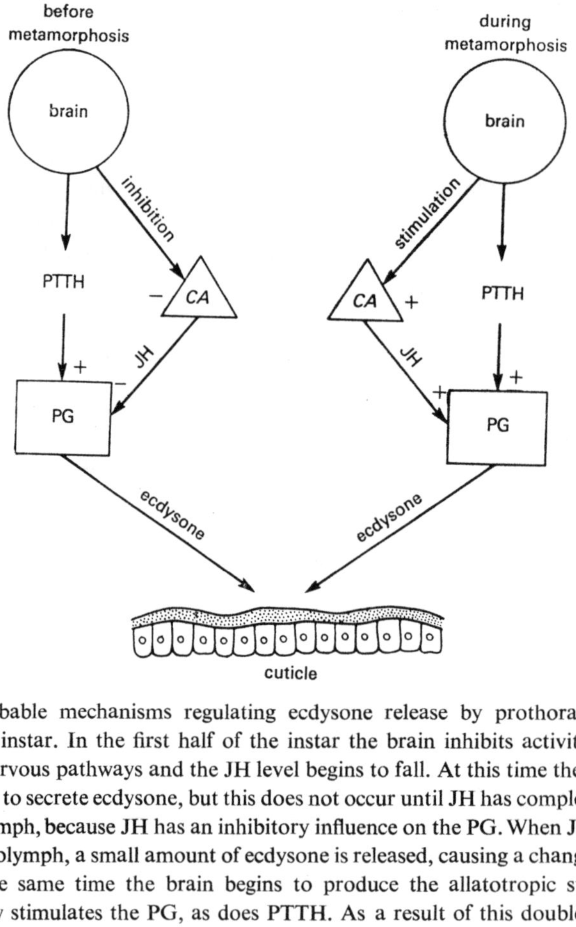

Figure 4.16. Probable mechanisms regulating ecdysone release by prothoracic glands (PG) in the last larval instar. In the first half of the instar the brain inhibits activity of the *corpora allata* (*CA*) via nervous pathways and the JH level begins to fall. At this time the brain is capable of stimulating PG to secrete ecdysone, but this does not occur until JH has completely disappeared from the haemolymph, because JH has an inhibitory influence on the PG. When JH has disappeared from the haemolymph, a small amount of ecdysone is released, causing a change of the insect's behaviour. At the same time the brain begins to produce the allatotropic substance. Newly produced JH now stimulates the PG, as does PTTH. As a result of this double stimulation the large quantity of ecdysone necessary for pupation is produced. After Cymborowski and Williams, unpublished data.

This can be demonstrated by placing ligatures behind the thorax and cutting off the head and thorax. The remaining part of the body then transforms into a pupal abdomen. At the time of the second ecdysone peak a small amount of juvenile hormone again appears in the haemolymph. As demonstrated in *Manduca sexta* larvae (Cymborowski and Williams, unpublished data), juvenile hormone has a double function at this time (Fig. 4.16). On the one hand, it stimulates the prothoracic glands to produce and release ecdysone, and on the other hand, it prevents the premature metamorphosis of some tissues. If the *corpora allata* are removed after the first period of ecdysone release (3rd day of the last instar) there is a considerable delay in pupation (depressed ecdysone level) and forms intermediate between the larva and pupa appear.

To sum up, the development and metamorphosis of insects depend on precise regulation of the levels of juvenile hormone and ecdysteroids. The latter, generally speaking, initiate moulting, whereas juvenile hormone determines the type of moult (larval or pupal). Pupation is preceded by complete disappearance of JH from the haemolymph, which allows the transition from larval to pupal development.

In the imago the prothoracic glands disappear, and their function, at least in some insects, is taken over by the ovaries. Juvenile hormone reappears and plays the role of a gonadotropic factor, as shown by Engelmann (1970) in studies on the cockroach (*Leucophaea maderae*). He demonstrated that the *corpora allata* are indispensable for normal maturation. This is so far the only evidence of the involvement of hormones in insect reproduction. Attempts to show that the gonads produce sex hormones in insects, as in vertebrates, have been unsuccessful.

The next chapter will be devoted to these problems.

4.4. Summary

In Figure 4.17 is shown a diagrammatic representation of the hormonal regulation of insect development and metamorphosis. Since the cuticle, or more strictly the epidermis, is the target tissue of the main insect hormones, studies to date have devoted most attention to the processes involved in moulting, i.e. shedding the old cuticle and producing the new one. In these processes three main hormones are involved: PTTH, ecdysone and juvenile hormone. PTTH is produced by the lateral neurosecretory cells of the brain and is released into the haemolymph by the *corpora cardiaca* (*CC*) and *corpora allata* (*CA*). The release of this hormone is synchronized by environmental conditions, especially the photoperiod.

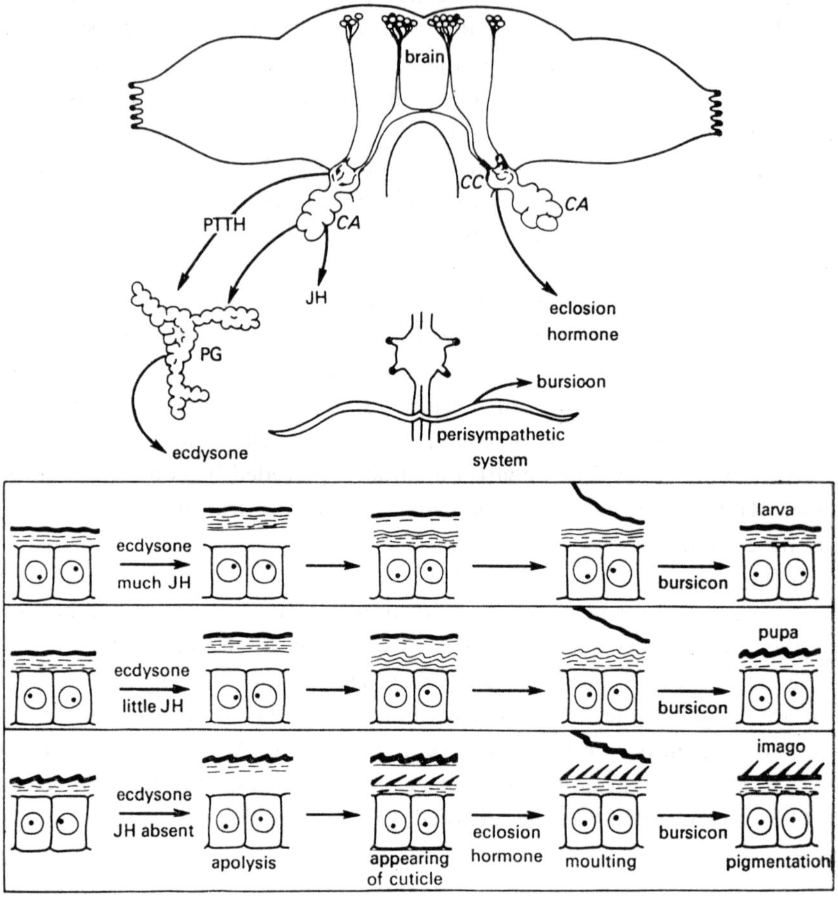

Figure 4.17. Functions of hormones in the development and metamorphosis of insects. After Riddiford and Truman (1978), with permission.

PTTH stimulates the prothoracic glands (PG) to produce and release ecdysone into the haemolymph. Ecdysone is further transformed, mainly by the fat body, to 20-hydroxyecdysone. This hormone acts on the epidermal cells, stimulating them to produce a new cuticle. When the new cuticle appears, the old one becomes detached as moulting fluid accumulates between the old and the new cuticle. This process is called apolysis. Taking place more or less at the same time are intensive DNA synthesis and epidermal cell division; the insect's dimensions increase. When the outer layer of the new cuticle is formed, digestive enzymes in the moulting fluid become activated, and digestion of the inner layers of the old cuticle starts.

The nature of the new cuticle (larval, pupal or imaginal) depends on the level of JH secreted by the *CA* during the period of 20-hydroxyecdysone activity. When the JH concentration in the larval haemolymph is high, larval cuticle is synthesized. However, when the activity of the *CA* diminishes and the JH concentration in the haemolymph falls, metamorphosis begins and pupal or imaginal cuticle is produced.

The final steps of the moulting process are controlled by two other hormones: eclosion hormone and bursicon. The shedding of the pupal cuticle is stimulated, at least in *Holometabola*, by eclosion hormone produced by the brain and released into the haemolymph by the *corpora cardiaca.* The cuticle of the newly formed insect is soft, enabling it to attain appropriate dimensions. During this period bursicon is released into the haemolymph, controlling the hardening and tanning of the cuticle. This hormone is produced by the neurosecretory cells of the abdominal ganglia and released via the perisympathetic system of the insect.

4.5. References

Agui N. 1975. Activation of prothoracic glands by brains *in vitro*. *J. Insect Physiol.*, **21**; 903–913.

Agui N., Granger N. A., Gilbert L. J., Bollenbacher W. E. 1979. Cellular localization of the insect prothoracicotropic hormone. *In vitro* assay of a single neurosecretory cell. *Proc. Nat. Acad. Sci. USA.*, **76**; 5694–5698.

Bergot B. J., Jamieson G. C., Ratcliff M. A., Schooley D. A. 1981. Identification of a new insect juvenile hormone. [In:] *Regulation of Insect Development and Behaviour*. F. Sehnal, A. Zabża, J. J. Menn and B. Cymborowski (eds.), pp. 173–182. Technical University Press, Wrocław.

Birkenbeil H., Gersch M. 1974. Weitere Untersuchungen zur hormonalen Beeinflüssung des Membranpotentials der Prothorakaldrüsenzellen von *Galleria mellonella. Zool. Jb.* (Physiol.,), **78**; 150–156.

Boguś M. J, Cymborowski B. 1981. Chilled *Galleria mellonella* larvae. Mechanism of supernumerary moultings. *Physiol. Entomol.*, **6**; 343–348.

Bollenbacher W. E., Vedeckis W. V., Gilbert L. J., O'Connor J. D. 1975. Ecdysone titres and prothoracic gland activity during the larval-pupal development of *Manduca sexta. Devel. Biol.*, **44**; 46–53.

Bouhniol J. J. 1936. Dans quelles limites l'écérebration des larves de Lépidoptères est-elle compatible avec leur nymphose. *C.R. Acad. Sci.* (*Paris*), **203**; 1182–1184.

Butenandt A., Karlson P. 1954. Über die Isolierung eines Metamorphose-Hormones der Insekten in Kristallisierter *Form. Z. Naturforsch.*, **9b**; 389–391.

Cherbas L., Cherbas P., Savakis C., Demetri G., Manteuffel-Cymborowska M., Yonge C. D., Williams C. M. 1980a. Studies of ecdysteroid action on a *Drosophila* cell line. [In:] *Invert-*

ebrate System—in vitro, pp. 217–228. Elsevier, North-Holland Biomedical Press, Amsterdam, New York, Oxford.

Cherbas P., Cherbas L., Demetri G., Manteuffel-Cymborowska M., Savakis C., Yonge C. D. Williams C. M. 1980b. Ecdysteroid hormone effects on a *Drosophila* cell line. [In:] *Gene Regulation by Steroid Hormones*. A. K. Roy and J. H. Clark (eds.), pp. 228–308. Springer-Verlag, New York, Heidelberg, Berlin.

Claret J. 1966. Mise en evidence du rôle photorécepteur du cerveau dans l'induction de la diapause chez *Pieris brassicae* (*Lepidoptera*). *Ann. Endocrinol.*, **27**; 311–320.

Clever U., Karlson P. 1960. Induktion von Puff-veränderungen in den Speicheldrüsen Chromosomen von *Chironomus tentans* durch Ecdyson. *Exp. Cell. Res.*, **20**; 623–626.

Cook D. J., Milligan J. V. 1972. Electrophysiology and histology of the medial neurosecretory cells in adult male cockroaches, *Periplaneta americana*. *J. Insect Physiol.*, **18**; 1197–1214.

Corbet S. A. 1971. Mandibular gland secretion of larvae of the flour moth *Anagasta kuehniella* contains an epideictic pheromone and elicits oviposition movements in a hymenopteran parasite. *Nature*, **232**; 481–484.

Cotter W. B. 1974. Social facilitation and development in *Ephestia kuehniella*. *Z. Science*, **183**; 747–748.

Cymborowski B., Giebułtowicz J. M. 1976. Effect of photoperiod on development and fecundity in the flour moth, *Ephestia kuehniella*. *J. Insect Physiol.*, **22**; 1213–1217.

Cymborowski B. 1977. Effect of lighting conditions on growth and developmental rhythms of singly reared *Dermestes vulpinus* (F.). *Com. Biochem. Physiol.*, **58A**; 99–101.

Cymborowski B., Boguś M. J. 1976. Juvenilizing effect of cooling on *Galleria mellonella*. *J. Insect Physiol.*, **22**; 669–672.

Cymborowski B., Riddiford L. M., Williams C. M., Beckage W. E. 1979. Endocrine control of starvation-induced supernumerary moultings in *Manduca sexta* larvae. *Western Regional Conference on Comp. Endocrinol. Ann. Soc. Zool.* Corvallis, Oregon.

Cymborowski B., Stolarz G. 1979. The role of juvenile hormone during larval-pupal transformation of *Spodoptera littoralis*. Switchover in the sensitivity of the prothoracic gland to juvenile hormone. *J. Insect Physiol.*, **25**; 939–942.

Cymborowski B. 1982. Receptory hormonów kontrolujących rozwój i metamorfozę owadów [Receptors of hormones controlling insect development and metamorphosis]. *Post. Biol. Kom.*, **9**; 53–66.

Delachambre J., Delbecque J-P., Provansal A., Grillot J. P., De Reggi M. L., Cailla H. L. 1979. Total and epidermal cyclic AMP levels related to variations of ecdysteroids and bursicon during metamorphosis of the mealworm, *Tenebrio mollitor*. *Insect Biochem.*, **9**; 95–100.

Engelman F. 1970. *The Physiology of Insect Reproduction*. Pergamon Press, Oxford.

Fain M. J., Riddiford L. M. 1975. Juvenile hormone titers in the hemolymph during late larval development of the tobacco hornworm, *Manduca sexta* (L.). *Biol. Bull.*, **149**; 506–521.

Ferkovich S. M., Silhaček D., Rutter R. 1975. Juvenile hormone binding proteins in the hemolymph of the indian meal moth. *Insect Biochem.*, **5**; 141–150.

Gabay S., Warburg M. R. 1976. Neurosecretory activity as related to feeding, mating, and oögenesis in the female cave tick, *Ornithodoros tholozani*. *J. Insect Physiol.*, **22**; 1291–1301.

Gersch M., Stürzebecker J. 1970. Experimentale Stimulierung der zellulären Aktivität der Prothorakaldrüsen von *Periplaneta americana* durch den Aktivations Factor. *J. Insect Physiol.*, **16**; 1813–1826.

Gersch M., Eibisch H. 1977. Synthesis of ecdysone-^{14}C and ecdysterone-^{14}C from cholesterol-^{14}C in cockroaches (*Periplaneta americana*) without molting glands. *Experientia*, **33**; 468.

Gibbs D., Riddiford L. M. 1977. Prothoracicotropic hormone in *Manduca sexta*: localization by a larval assay. *J. exp. Biol.*, **66**; 255–266.

Giebułtowicz J. M., Ždarek J., Chróścikowska U. 1980. Cocoon spinning behavior in *Ephestia kuehniella*; Correlation with endocrine events. *J. Insect Physiol.*, **26**; 459–464.

Gilbert L. J., Bollenbacher W. E., Granger N. A. 1980. Insect endocrinology: Regulation of endocrine glands, hormone titer, and hormone metabolism. *Ann. Rev. Physiol.*, **42**; 493–510.

Girardie A., De Reggi M. 1978. Moulting and ecdysone release in response to electrical stimulation of protocerebral neurosecretory cells in *Locusta migratoria. J. Insect Physiol.*, **24**; 797–802.

Goodman W., Gilbert L. J. 1978. The hemolymph titer of juvenile hormone-binding protein and binding sites during the fourth larval instar of *Manduca sexta. Gen. Comp. Endocrinol.*, **35**; 27–43.

Granger N. A., Sehnal F. 1974. Regulation of larval *corpora allata* in *Galleria mellonella. Nature*, **251**; 415–417.

Hagedorn H. H., Shapiro J. P., Hanaoka K. 1979. Ovarian ecdysone secretion is controlled by a brain hormone in an adult mosquito? *Nature*, **282**; 92–94.

Herman W. S. 1967. The ecdysial glands of arthropods. *Int. Rev. Cytol.*, **22**; 269–347.

Highnam K. C., Hill L. 1977. *Comparative Endocrinology of the Invertebrates*. Second Edition, pp. 104–130. Edward Arnold, London.

Hikino H., Takemoto T. 1974. Ecdysones of plant origin. [In:] *Invertebrate Endocrinology and Hormonal Heterophylly*. W. J. Burdette (ed.), pp. 185–203. Springer-Verlag, New York, Heidelberg, Berlin.

Horn D. H. S., Middleton E. J., Wunderlich J. A., Hampshire F. 1966. Identity of the molting hormones of insects and crustaceans. *Chem. Commun.*, 339–341.

Hsiao T. H., Hsiao C. 1979. Ecdysteroids in the ovary and the egg of the greater wax moth. *J. Insect Physiol.*, **25**; 45–52.

Hüber R., Hoppe W. 1966. Zur Chemie des Ecdysons. VII. Die Kristall- und Molekulstrukturanalyse des Insekteverpuppungshormons Ecdyson mit der Automatisierten Faltmolekülmethode. *Chem. Ber.*, **98**; 2403–2424.

Ishizaki H., Ichikawa M. 1967. Purification of the brain hormone of the silkworm, *Bombyx mori. Biol. Bull.*, **133**; 355–368.

Johnston W. S., McFarlane J. E. 1973. Occurrence of the group effect in *Acheta domesticus* (L.) (*Orthoptera*; *Gryllidae*) in relation to parental age. *Ann. entomol. Soc. Québec*, **18**; 61–65.

Judy K. J., Schooley D. S., Dunham L. L., Hall M. S., Bergot B. J., Siddall J. B. 1973. Isolation, structure, and absolute configuration of a new natural insect juvenile hormone from *Manduca sexta. Proc. Nat. Acad. Sci. USA.*, **70**; 1509–1513.

Karlson P., Hofmeister H., Hummel H., Hocks P., Spiteller G. 1965. Zur Chemie des Ecdysons. VI. Reaktionen des Ecdysonmoleküls. *Chem. Ber.*, **98**; 2394–2402.

Karlson P., Bode C. 1969. Die Inaktivierung des Ecdysons bei der Schmeiss-Fliege, *Calliphora erythrocephala. J. Insect Physiol.*, **165**; 111–118.

Karlson P., Koolman J. 1976. Ecdysone mode of action, inactivation and elimination. *Coll. internat. C.N.R.S.*, **251**; 325–341.

King D. S., Bollenbacher W. E., Borst D. W., Vedeckis W. V., O'Connor J. D., Ittycheriah P. J., Gilbert L. J. 1974. The secretion of α-ecdysone by the prothoracic glands of *Manduca sexta in vitro*. *Proc. Nat. Acad. Sci. USA.*, **71**; 793–796.

Koolman J., Hoffmann J. A., Dreyer M. 1975. Moulting hormone in *Locusta migratoria*: rate of excretion during the last larval instar. *Experientia*, **31**; 247–249.

Kryspin J., Dutkowski A. B., Cymborowski B. 1975. The influence of illumination conditions on growth and development of *Galleria mellonella*. *Bull. Acad. Pol. Sci.*, **22**; 803–808.

Lafont R., Delbeque J. P., De Hys L., Mauchamp B., Pennetier J. L. 1974. α- and β-ecdysone levels in insect haemolymph, correlation with development events. *Experientia*, **31**; 1241–1242.

Lees A. D. 1960. Some aspects of animal photoperiodism. *Cold Spring Harb. Symp. Quant. Biol.*, **25**; 261–268.

Lees A. D. 1964. The location of the photoperiodic receptors in the aphid *Megoura viciae* Buckton. *J. Exp. Biol.*, **41**; 119–133.

Lockshin R. A., Beaulaton J. 1974. Programmed cell death. *Life Sci.*, **15**; 1549–1565.

Maroy P., Dennis R., Beckeris C., Sage B. A., O'Connor J. D. 1978. Demonstration of an ecdysteroid receptor in a cultured cell line of *Drosophila melanogaster*. *Proc. Nat. Acad. Sci. USA.*, **75**; 6035–6038.

Mayer R. T., Svoboda J. A., Weirich G. F. 1978. Ecdysone 20-hydroxylase in midgut mitochondria of *Manduca sexta* (L.). *Z. Physiol. Chem.*, **359**; 1247–1257.

McDaniel C. N., Berey S. J. 1967. Activation of the prothoracic glands of *Antheraea polyphemus*. *Nature*, **214**; 1032–1034.

Morris G. P., Steel C. G. H. 1977. Sequence of ultrastructural changes induced by activation in the posterior neurosecretory cells in the brain of *Rhodnius prolixus* with special reference to the role of lysosome. *Tissue. Cell.*, **9**; 547–562.

Muszyńska-Pytel M., Cymborowski B. 1978. The role of serotonin in regulation of the circadian rhythms of locomotor activity in the cricket (*Acheta domesticus* L.). II. Distribution of serotonin and variations in different brain structures. *Comp. Biochem. Physiol.*, **59 C**; 17–20.

Muszyńska-Pytel M. 1987. Invertebrate System *in vitro*. Elsevier, Norh-Holland Biomedical Press, Amsterdam, New York, Oxford.

Nakanishi K., Koreeda M., Sasaki S., Chang M. L., Hsu M. Y. 1966. Insect hormones. I. The structure of ponasterone A, from insect-moulting hormone from the leaves of *Podocarpus nakii* Hay. *Chem. Commun.*, 915–917.

Nagasawa H., Isogai A., Suzuki A., Tamura S., Ishizaki H. 1979. Purification and properties of the prothoracicotropic hormone of the silkworm, *Bombyx mori*. *Develop. Growth Diff.*, **21**; 29–38.

Nijhout H. F., Williams C. M. 1974. Control of moulting and metamorphosis in the tobacco hornworm, *Manduca sexta* (L.): Growth of the last-instar larva and the decision to pupate. *J. Exp. Biol.*, **61**; 481–491.

Nijhout H. F. 1975. A threshold size for metamorphosis in the tobacco hornworm, *Manduca sexta* (L.). *Biol. Bull.*, **149**; 214–225.

Nijhout H. F. 1979. Stretch-induced moulting in *Oncopeltus fasciatus*. *J. Insect Physiol.*, **25**; 277–281.

Oberlander H., Berry S. J., Krishnakumaran A., Schneiderman H. A. 1965. RNA and DNA synthesis during activation and secretion of the prothoracic glands of saturniid moths. *J. exp. Zool.*, **159**; 15–32.

Ohtaki T. 1966. On the delayed pupation of the fleshfly, *Sarcophaga peregrina* Robineau-Desvoidy. *Japan J. Med. Sci. Biol.*, **19**; 97–104.

Ohtaki T., Milkman R. D., Williams C. M. 1968. Dynamics of ecdysone secretion and action in the fleshfly *Sarcophaga peregrina.* Biol. Bull., **135**; 322–334.

Piepho H. 1940. Über die Hemmung der Verpuppung durch *corpora allata.* Untersuchungen an der Wachsmothe *Galleria mellonella* L. *Biol. Zbl.*, **60**; 367–393.

Reibstein D., Law J. H. 1973. Enzymatic synthesis of insect juvenile hormones. *Biochem. Biophys. Res. Commun.*, **55**; 266–272.

Reynolds S. E. 1977. Control of cuticle extensibility in the wings of adult *Manduca* at the time of eclosion: effect of eclosion hormone and bursicon. *J. Exp. Biol.*, **70**; 27–39.

Reynolds S. E., Taghert P. H., Truman J. W. 1979. Eclosion hormone and bursicon titer and the onset of hormonal responsiveness during the last day of adult development in *Manduca sexta* (L.). *J. Exp. Biol.*, **78**; 77–86.

Reynolds S. E., Truman J. W. 1979. Eclosion hormone. [In:] *Insect (Peptide) Neurohormones.* T. Miller (ed.). Springer-Verlag, New York, Heidelberg, Berlin.

Riddiford L. M. 1978. Ecdysone-induced change in cellular commitment of the epidermis of the tobacco hornworm, *Manduca sexta*, at the initiation of metamorphosis. *Gen. Comp. Endocrinol.*, **34**; 438–446.

Riddiford L. M., Truman J. W. 1978. Biochemistry of insect hormones and insect growth regulators. [In:] *Biochemistry of Insects.* M. Rockstein (ed.), pp. 307–357. Academic Press, New York, San Francisco, London.

Riddiford L. M. 1980a. Interaction of ecdysteroids and juvenile hormone in the regulation of larval growth and metamorphosis of the tobacco hornworm. [In:] *Progress in Ecdysone Research.* J. A. Hoffman (ed.). Elsevier, North-Holland, Amsterdam, New York, Oxford.

Riddiford L. M. 1980b. Prothoracicotropic hormone-secreting cells in the insect brain. *Nature*, **285**; 6.

Röller H., Dahm K. H., Sweeley C. C., Trost B. M. 1967. Die Struktur des Juvenilhormons. *Angew. Chem.*, **79**; 190–191.

Rudnicka M., Sehnal F., Jarolim V., Kochman M. 1979. Hydrolysis and binding of the juvenile hormone in the hemolymph of *Galleria mellonella. Insect Biochem.*, **9**; 569–575.

Safranek L., Cymborowski B., Williams C. M. 1980. Effect of juvenile hormone on ecdysone-dependent development in the tobacco hornworm, *Manduca sexta. Biol. Bull.*, **158**; 248–256.

Scharrer B., Würzelmann S. 1978. Neurosecretion. XVII. Experimentally induced release of neurosecretory material by exocytosis in the insect *Leucophaea maderae. Cell Tissue Res.*, **190**; 173–180.

Schmialek P. 1973. Ribonucleoprotein particle in epidermis cells as the receptor for juvenile hormone. *Nature*, **245**; 267–268.

Schmialek P., Geyer A., Miosga V., Nündel M., Zapf B. 1975. Juvenil hormonbindende Substanzen mit allosterischen Eigenschaften in den Ovarien von *Tenebrio mollitor* L. *Z. Naturforsch.*, **30c**; 730–733.

Sehnal F., Granger N. A. 1975. Control of *corpora allata* function in larvae of *Galleria mellonella. Biol. Bull.*, **148**; 106–116.

Sehnal F., Zabża A., Menn J. J., Cymborowski B. (eds.). 1981. *Regulation of Insect Development and Behaviour*. Technical University Press, Wrocław.

Seligman J. M. 1980. Bursicon. [In:] *Neurohormonal Techniques in Insects*. T. A. Miller (ed.), pp. 137–153. Springer-Verlag, New York, Heidelberg, Berlin.

Shaaya E., Karlson P. 1965. Der Ecdysontiter während der Insektenentwicklung. IV. Die Entwicklung der Lepidopteren *Bombyx mori* L. und *Cerura vinula* L. Dev. Biol., **11**; 424–432.

Shaaya E. 1969. Untersuchungen über die Verteilung des Ecdysons in verschiedenen Gewebe von *Calliphora erythrocephala* und über seine biologische Halbwertzeit. *Z. Naturf.*, **24b**; 718–721.

Sobczyk J., Cymborowski B., Węsierska M. 1979. Effect of density on developmental rhythms in *Dermestes vulpinus*. *Bull. Acad. Pol. Sci.*, **27**; 1053–1057.

Steel C. G. H., Lees A. D. 1977. The role of neurosecretion in the photoperiodic control of polymorphism in the aphid *Megoura viciae*. *J. exp. Biol.*, **67**; 117–135.

Tanaka Y. 1950. Studies on hibernation with special reference to photoperiodicity and breeding of the chinese tussar silkworm—III. *J. Sevic. Sci. Japan.*, **19**; 580–590.

Taylor H. M. Truman J. W. 1974. Metamorphosis of the abdominal ganglia of the tobacco hornworm, *Manduca sexta*: changes in populations of identified motor neurons. *J. Comp. Physiol.*, **90**; 367–388.

Truman J. W. 1972. Physiology of insect rhythms. I. Circadian organization of the endocrine events underlying the moulting cycle of larval tobacco hornworms. *J. Exp. Biol.*, **57**; 805–820.

Truman J. W. 1973. Physiology of insect ecdysis. III. Relationship between the hormonal control of eclosion and of tanning in the tobacco hornworm, *Manduca sexta*. *J. Exp. Biol.*, **58**; 821–829.

Truman J. W., Riddiford L. M. 1974. Physiology of insect rhythms. III. The temporal organization of the endocrine events underlying pupation of the tobacco hornworm. *J. Exp. Biol.*, **60**; 371–382.

Truman J. W. 1976. External photoreception in insects. *Photochem. Photobiol.*, **23**; 215–225.

Truman J. W., Fallon A. M., Wyatt G. R. 1976. Hormonal release of programmed behavior in silk moth: probable mediation by cyclic AMP. *Science*, **194**; 1432–1434.

Truman J. W. 1978. How moths "turn on": a study of the action of hormones on the nervous system. *Amer. Sci.*, **61**; 700–706.

Truman J. W. 1978. Hormonal release of stereotyped motor programmes from the isolated nervous system of the *Cecropia* silkmoth. *J. Exp. Biol.*, **74**; 151–173.

Vedeckis W. V., Gilbert L. J. 1973. Production of cyclic AMP and adenosine by the brain and prothoracic glands of *Manduca sexta*. J. Insect Physiol., **19**; 2445–2457.

Vincent J. F. V. 1972. The dynamics of release and the possible identity of bursicon in *Locusta migratoria migratorioides*. *J. Insect Physiol.*, **18**; 757–780.

Wartoń S. S., Dutkowski A. B. 1977. Ultrastructure of the neurosecretory cells of *pars intercerebralis* of *Galleria mellonella* (*Lepidoptera*) after noradrenaline administration. *Gen. Comp. Endocrinol.*, **33**; 179–186.

Wartoń S. S., Dutkowski A. B. 1978. Ultrastructural analysis of the action of reserpine on the brain neuroendocrine system of the wax moth, *Galleria mellonella* L. (*Lepidoptera*). *Cell Tissue Res.*, **192**; 143–155.

Wigglesworth V. B. 1934. The physiology of ecdysis in *Rhodnius*. II. Factors controlling moulting and metamorphosis. *Quart. J. Micr. Sci.*, **77**; 191–222.

Williams C. M. 1946. Physiology of insect diapause: the role of the brain in the production and termination of pupal dormancy in the giant silkworm *Platysamia cecropia*. *Biol. Bull.*, **90**; 234–243.

Williams C. M. 1956. The juvenile hormone of insects. *Nature*, **178**; 212–213.

Williams C. M., Adkisson P. L. 1964. Physiology of insect diapause. XIV. An endocrine mechanism for the photoperiodic control of pupal diapause in the oak silkworm, *Antheraea pernyi*. *Biol. Bull.* Woods Hole, **127**; 511–525.

Williams C. M., Kafatos F. C. 1971. Theoretical aspects of the action of juvenile hormone. *Bull. Soc. Entomol. Suisse.*, **44**; 151–162.

Williams C. M. 1976. Juvenile hormone... in retrospect and in prospect. [In:] *The Juvenile Hormones*. L. J. Gilbert (ed.), pp. 1–14. Plenum Press, New York, London.

Woolever P., Pipa R. L. 1970. Spatial and feeding requirements for pupation of last instar larvae of *Galleria mellonella*. *J. Insect Physiol.*, **16**; 251–262.

Yamazaki M., Kobayashi M. 1969. Purification of the proteinic brain hormone of the silkworm, *Bombyx mori*. *J. Insect Physiol.*, **15**; 1981–1990.

Yound M. A., King D. S., Fristrom J. W. 1978. Ecdysteroid receptors in imaginal discs of *Drosophila melanogaster*. *Proc. Nat. Acad. Sci. USA.*, **75**; 6039–6043.

Yound M. A. 1980. Receptors for 20-hydroxyecdysone in imaginal discs of *Drosophila*. [In:] *Receptors for Neurotransmitters, Hormones and Pheromones in Insects*. D. B. Satelle *et al.* (eds.), pp. 233–243. Elsevier, North-Holland Biomedical Press.

Zimmerman W. F., Goldsmith T. H. 1971. Photosensitivity of the circadian rhythm and of visual receptors in carotenoid-depleted *Drosophila*. *Science*, **171**; 1167–1168.

Chapter 5

Hormonal regulation of insect reproduction

Insects belong to a nontypical group of animals in which the same hormones regulate growth and metamorphosis and control reproduction during adult life. The *corpora allata* (*CA*) resume their function in the adult, producing JH which controls oogenesis (de Wilde and De Loof, 1973) and the maturation of the male's accessory sex glands (Leopold, 1976), and participates in the regulation of reproductive behaviour (de Wilde, 1964; Truman and Riddiford, 1977). The prothoracic glands usually degenerate during pupal development; great interest was therefore aroused by the discovery that in some insects the ovaries act as endocrine glands, producing ecdysone and releasing it into the haemolymph and directly into the developing eggs, where it may control oogenesis or embryonic development (Gilbert *et al.*, 1980).

The present chapter deals with some aspects of hormonal regulation of insect reproduction, with the omission, however, of pheromones. As ectohormones, these are beyond the scope of insect endocrinology. The influence of hormones, particularly JH, on the production and release of pheromones is mentioned briefly.

5.1. Characteristics of the reproductive system

Insects are dioecious animals, although the parthenogenetic type of reproduction also occurs (e.g. many aphid species). Contrary to the situation in vertebrates, however, in a great majority of insects the secondary sexual characters are not controlled by hormones produced by the gonads. This was already evident in the first studies of Stefan Kopeć in the years 1908–1912. He found that removal or transplantation of the gonads has no influence on the development of the secondary sexual characters (Kopeć, 1912), as is the case in vertebrates. In the literature there is at least one piece of evidence that both primary and secondary sexual characters are induced by the "androgenic" hormone derived from the paired androgenic glands closely associated anatomically with the male gonads. As demonstrated by Naisse (1965), transplantation of the testes of the glow-worm (*Lampyris noctiluca*) together with the associated androgenic glands to a larval female results in the transformation of the larva to a male. On the other hand, transplantation of ovaries to male larvae has no effect. It was also found that the later the testes are transplanted to females the weaker is the effect. Transplantation during the pupal period has a very slight masculinizing effect; in adults the effect is practically undetectable.

The formation and maturation of the gonads are part of metamorphosis. Gonads arise from pits in the embryonic ectoderm which gradually differentiate during the course of development and metamorphosis. From the theoretical point of view, therefore, the development and function of the gonads may be controlled by the same hormones that regulate metamorphosis. As we shall see later, this is actually the case.

5.2. Female gonads

In most insects the ovaries are paired organs (Fig. 5.1), each of which is surrounded by a thin epithelium under which lie connective tissue and smooth muscle. The number of ovarioles in each ovary varies among insect species. In the termite queen there may be ten thousand or more. At the top of the ovariole there is a germarium where the oocytes differentiate. Further along the ovariole lies the vitellarium in which yolk is transferred to the oocytes. This process is named vitellogenesis. Each ovariole is connected with the oviduct, which carries the mature eggs to the exterior.

Two main types of ovarioles can be distinguished, based on the manner in which the oocytes differentiate and are supplied with nutritients during their early development (Fig. 5.2). The ovarioles may be panoistic or meroistic; the latter may be divided into polytrophic and telotrophic types. In panoistic ovarioles the germarium contains only oogonia which become surrounded by mesodermal cells as they move down the ovariole, and it is in this way that the follicular vesicle arises. This type of ovariole is found in *Odonata*, *Ephemeroptera*, *Orthoptera*, *Plecoptera* and in several other orders of insects. In polytrophic ovarioles, the follicular vesicle contains a number of nutritive cells in contact with the oocyte. This type of ovariole occurs in *Anoplura*, *Dermaptera*, *Trichoptera*, *Lepidoptera*, *Diptera*, *Hymenoptera* and other orders. Polytrophic and telotrophic ovarioles

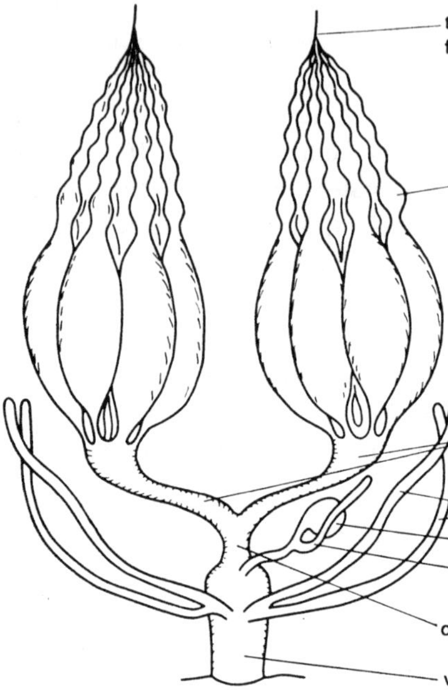

Figure 5.1. Structure of insect ovaries. Each ovary contains a number of ovarioles connected to the oviduct. Usually, this organ is also connected to the spermatheca in which spermatozoa accumulate during copulation. Accessory sex glands produce secretion which covers the forming egg, facilitating its emergence and protecting it. After Highnam and Hill (1977), with permission.

have been described in the *Coleoptera* by several authors (Telfer, 1975; Kloc and Matuszewski, 1977).

In one species of *Coleoptera*, *Steraspis speciosa*, a new type of ovariole has been identified (Martoja, 1964). The germarium and vitellarium together occupy only a small part of the ovariole (see Fig. 5.2), under which extends a long glandular structure. The role of this structure is to secrete a coating for the egg. A trophic region has not yet been detected in this type of ovariole. It would seem that this is a modified type of telotrophic ovariole, and the term "adenotrophic" is suggested for it.

Another type of ovariole has been described by Büning (1979) in *Sialis flavilatera* (*Megaloptera*). In this species the germarium consists of two distinct types of tissue. The middle part is occupied by nurse cells, whose nuclei lie in a common cytoplasm (syncytium) rich in ribosomes and mitochondria. The

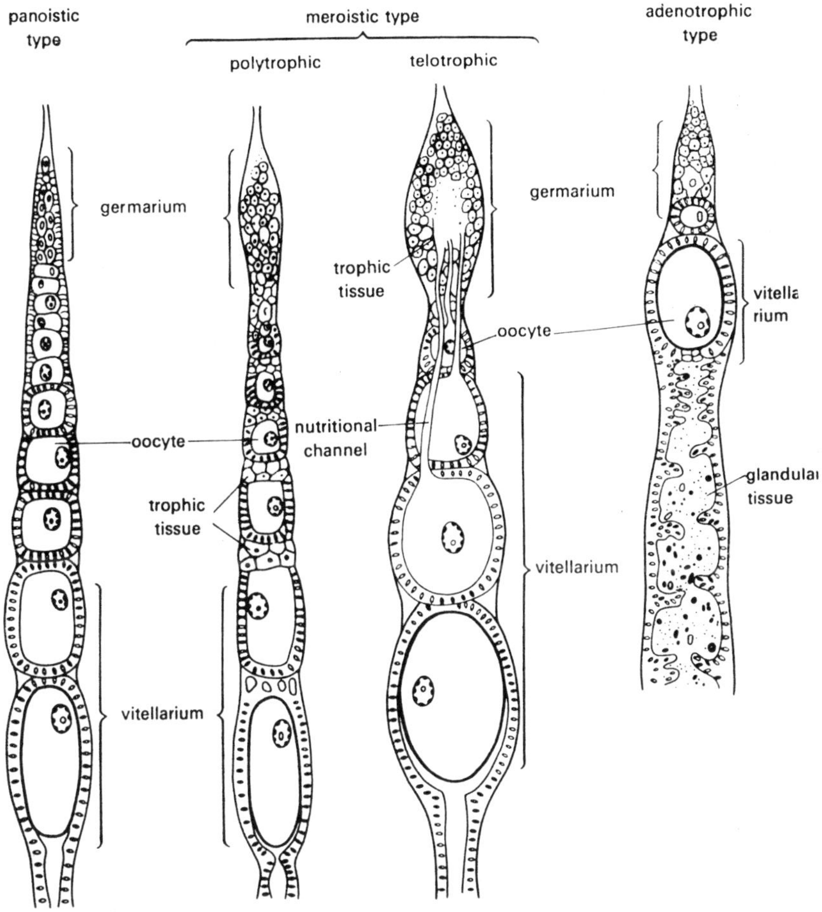

Figure 5.2. Four types of ovarioles occurring in insects. Details in text. After Engelmann (1970).

syncytium is surrounded by a single layer of nurse cells. Each of these surrounding nurse cells is connected with the syncytium by a protoplasmic bridge which plays a role in nutritient supply during differentiation of the oocyte.

Accessory sex glands are associated with the ovaries. At least four types of connection between the accessory sex glands and the ovaries have been distinguished. In *Dictyoptera* they are connected with the oviduct, in *Acrididae* with the anterior part of the oviduct, and in *Locusta migratoria* the whole oviduct effectively functions as an accessory sex gland. The secretion of these glands coats the developing eggs, protecting them from injury during and after oviposition.

5.3. Male gonads

Male gonads are usually paired (Fig. 5.3), although in many species they have coalesced into a single organ. The most distal part of each testis is occupied by the germaria, in which spermatogenesis takes place. Maturation of spermatids may occur in various parts of the testes, frequently distant from the germaria. Paired canaliculi (*vasa deferentia*) run from the testes, carrying the semen to the seminal vesicles where it is temporarily stored. Later, during copulation, the spermatozoa pass into the common ejaculatory channel (*ductus ejaculatorius*).

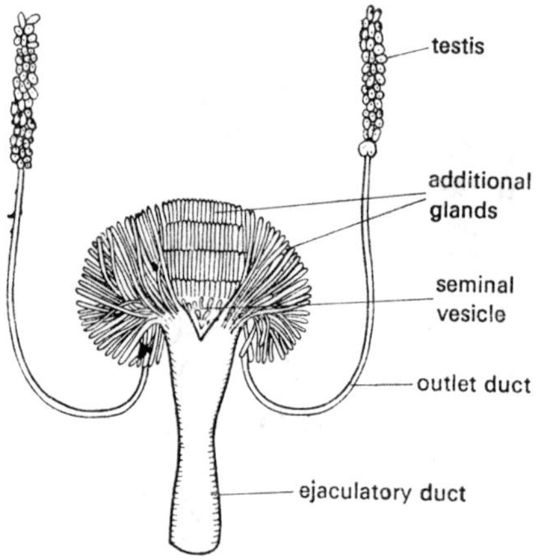

Figure 5.3. Structure of reproductive system of male cockroach. Accessory sex glands are also called colleteral glands. Details in text. After Highnam and Hill (1971), with permission.

5.4. Hormonal regulation of egg maturation

5.4.1. The role of neurosecretory cells

Thomsen (1948) demonstrated for the first time that destruction of the medial neurosecretory cells in the brain of *Calliphora erythrocephala* completely blocks egg maturation. Reimplantation of these cells restores maturation, at least in some insects. Removal of the *corpora allata* does not prevent egg maturation in this species. On the basis of these experiments it was proposed that the neurosecretory cells of the *pars intercerebralis* produce a gonadotropic hormone. Similar results were obtained in the mosquito *Aedes taeniorhynchus*, which did not produce eggs after destruction of the *pars intercerebralis* (Lea, 1967). In this species, however, it appeared that egg maturation depends on the activation of the *CC*, since maturation is blocked by their removal. The complexity of the process of egg maturation is demonstrated by the experiments of Wilkens (1968), in which maturation was blocked by destruction of the *pars intercerebralis* in *Sarcophaga bullata*, but it was not resumed after implantation of active *CC*. Similar results were obtained in experiments with the locust (Lazaroviçi and Pener, 1978). The hypothesis was therefore advanced that protein synthesis stimulated by JH is indispensable for egg maturation, whereas the neurosecretory factor acts as a true gonadotropic hormone, stimulating the transport of the proteins to the oocyte.

Some investigators have suggested that neurosecretion is involved in the regulation of oviposition in various insect species. Enders (1956), for instance, found that under *in vitro* conditions brain extracts increase the frequency of contraction of the oviducts in the stick insect (*Carausius morosus*), and this may be important in egg-laying under normal conditions. On the other hand, Nayar (1958) demonstrated that when neurosecretory cells taken from the brain of *Iphita limbata* just before oviposition are transplanted into the body cavity of a female with immature eggs, the recipient female lays her eggs. Nayar suggests the following mechanism of action of these cells: fully mature eggs secrete a hormone stimulating the neurosecretory cells of the brain to release its secretory material; by increasing the frequency of ovarian contractions, this material accelerates egg laying. Such a mechanism, however, has not been found so far in other insects.

Results of studies on the cricket (*Acheta domesticus*) suggest the participation of neurosecretory cells of both the brain and the suboesophageal ganglion in the regulation of RNA synthesis in the follicular epithelium of the oocytes (Dutkowski and Cymborowski, 1970, 1971, 1973). RNA synthesis is inhibited by the neurosecretion from the suboesophageal ganglion; incubation of oocytes *in vitro* in the presence of the *corpora allata*, however, stimulates synthesis.

Although frequently controversial, these data seem to suggest that neurosecretion from the *pars intercerebralis* regulates certain stages of the reproductive process in insects. It can be assumed, however, that the involvement of these cells may differ among species. In the extreme case they may not participate at all in reproduction, as has been observed in *Rhodnius prolixus* (Wigglesworth, 1936) and in the cockroach *Leucophaea maderae* (Engelmann, 1964). Moreover, the results of studies on the effect of cauterization of the neurosecretory cells on egg maturation are extremely difficult to interpret, because this treatment greatly affects feeding in these insects (Engelmann, 1965). It is known that normal feeding is essential for egg maturation in most insects, with the exception of those which do not feed after the imaginal moult (e.g. *Saturnidae*).

5.4.2. Role of JH in egg maturation

The involvement of JH in the process of egg maturation was observed for the first time by Wigglesworth (1936) in *Rhodnius prolixus*. In this species the eggs mature after the female has fed on blood. If, immediately after the blood meal, the brain is removed but the *corpora allata* are left in place, maturation of the eggs is not inhibited, whereas if the brain is removed together with the *CA* egg maturation does not occur (Fig. 5.4). These experiments clearly show that the *CA*, which cease to function before metamorphosis (see Chapter 4), resume JH production in the adult, and the hormone then stimulates egg maturation. There are now abundant data indicating that JH acts as a gonadotropic hormone in adult insects (Engelmann, 1970; Highnam and Hill, 1977).

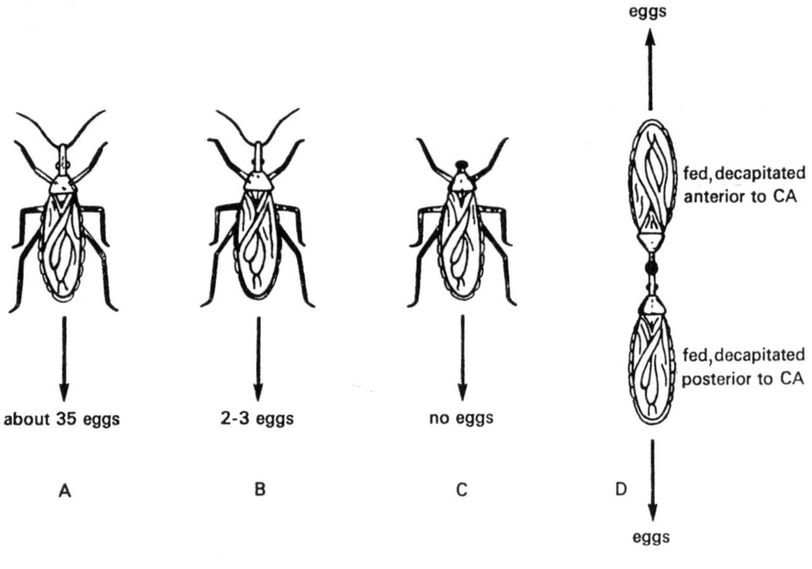

Figure 5.4. Illustration of experiments performed by Wigglesworth (1936) on *Rhodnius prolixus*, indicating the role of *corpora allata* in egg production.
A—insect normally produces about 35 eggs. Starved insect (B) lays only a few eggs. Normally fed insect, decapitated so as to remove *corpora allata* (*C*), does not produce eggs; fed insect with head and *corpora allata* removed, connected to decapitated insect with *corpora allata* intact (D)—both produce eggs.

In many insects the terminal oocytes develop synchronously in all the ovarioles, and all mature eggs are deposited simultaneously. Their place in the ovariole is taken by the next most mature eggs, and the cycle is repeated. As shown by the investigations of Stranways-Dixon (1961), changes in oocyte size are accompanied by similar changes in the *corpora allata* (Fig. 5.5). The dimensions of the *CA* in *Calliphora erythrocephala* increase in the initial stages of vitellogenesis, and

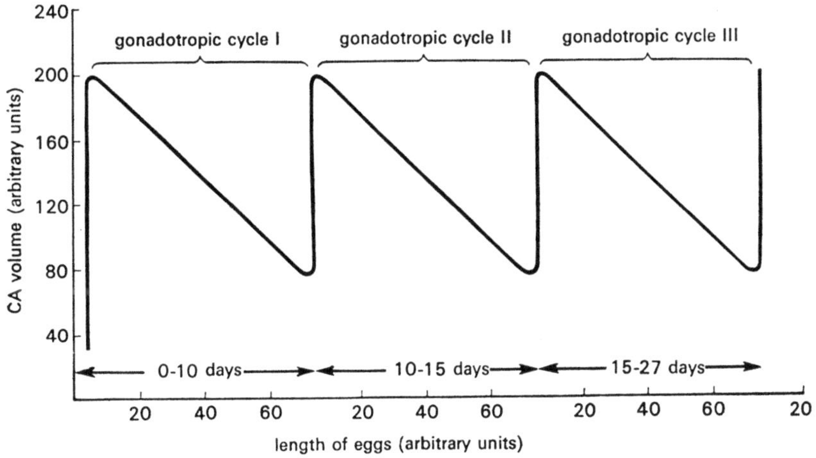

Figure 5.5. Cyclic changes in size of *corpora allata* in *Calliphora erythrocephala* during the gonadotropic cycle. Description in text. After Strangways-Dixon (1961).

then diminish with increasing oocyte size. The dimensions of the *CA* are least shortly before oviposition. This is evidence of the involvement of the *CA* in the process of egg maturation. The oviposition of one batch of eggs and the passage of the next one into the oviduct seem to be under nervous control, as indicated by studies on the stick insect (Thomas, 1979a, b; Mesnier, 1980). Artificial arrest of oviposition inhibits maturation of further eggs without intervention by the endocrine system, indicating the involvement of the nervous system in this process.

Adams *et al.* (1968) suggest that in the house fly (*Musca domestica*) an oostatic hormone is produced and released into the haemolymph after maturation of a batch of eggs in the ovaries. The presence of this hormone inhibits the development of the next batch until the earlier eggs are deposited. This hormone may act by inhibiting the activity of the *corpora allata*, which would explain the cyclic changes in the size of these organs in many of the insect species examined to date.

Engelmann (1964) suggested that in *Leucophaea maderae* the mechanism is quite different. According to this author, the ootheca begins to produce an unknown substance whilst it contains mature eggs, this substance acting in turn on neurons in the brain to inhibit the activity of the *corpora allata*. Further research will prove which of these hypotheses is true. Differences in the mechanisms of activation and inhibition of *CA* function may be expected among insect species.

The juvenile hormone influences three essential processes associated with egg maturation (Riddiford and Truman, 1978). These are (1) growth of the oocytes, (2) stimulation of synthesis of special proteins (vitellogenins) by the fat body, and (3) regulation of the absorption of these proteins by the oocytes (Fig. 5.6).

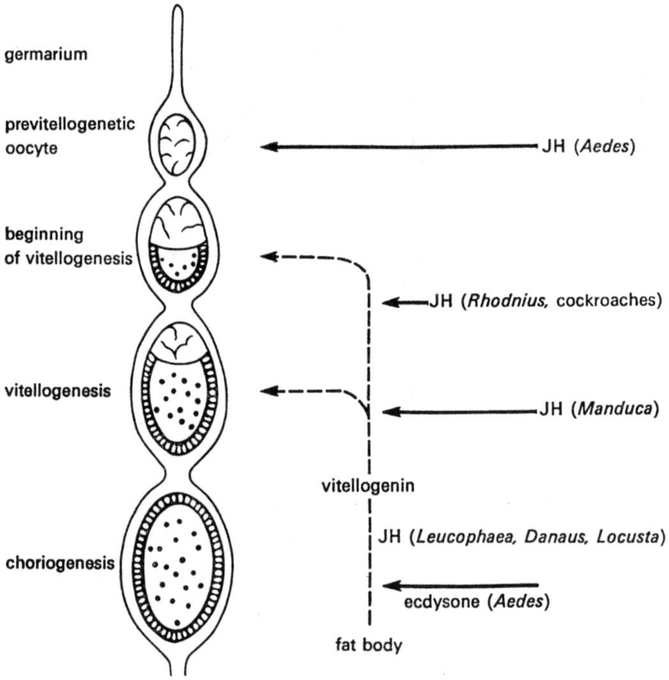

Figure 5.6. Hormonal regulation of particular stages of oogenesis. Description in text. After Riddiford and Truman (1978), with permission.

The first stages of oocyte growth occur in the germarium of the ovaries during metamorphosis, and are therefore under the influence of ecdysone but not of JH. In most insects, however, the oocytes begin growth after the imaginal moult, when secretion of JH into the haemolymph is resumed. In the mosquito

Aedes aegypti, for instance, secretion of JH immediately after eclosion of the imago from the pupa leads to an increase in size of the oocytes from 80 to 110 μm; the oocytes then enter the so-called resting state until the insect takes a blood meal. This initiates the process of vitellogenesis (Lea, 1963). In *Manduca sexta*, however, there is a relatively long interval between the imaginal moult and the secretion of JH, and the oocytes do not increase in size until the hormone appears in the haemolymph.

It is necessary that the oocytes attain an appropriate size in order that vitellogenesis can begin. Telfer (1954) was the first to demonstrate that the haemolymph of *Hyalophora cecropia* females contains proteins which are not found in that of males. It appeared later that these proteins accumulate in maturating eggs, and are generally found in females of various insect species; they were designated vitellogenins. They are ordinary proteins with a very high molecular weight, of the order of 200 000 to 500 000. Further investigations showed that the synthesis of these proteins is regulated by JH (Engelmann, 1970, 1979). In *Locusta migratoria* JH is necessary for the maturation of the fat body, enabling it to produce vitellogenins (Chen *et al.*, 1979; Couble *et al.*, 1979). Vitellogenin synthesis does not occur in females from which the *corpora allata* have been removed. Synthesis can be stimulated by administration of JH or its analogues, and usually occurs about 48 h after treatment. Receptors for JH have been detected in the fat body of *Locusta migratoria*, suggesting that the hormone can directly influence the synthesis of specific mRNA in the fat body (Riddiford, 1980). If this is the case (the results have yet to be confirmed in *in vitro* studies), this mechanism would resemble that observed in vertebrates, in which oestrogen stimulates vitellogenin synthesis in the liver.

In *Rhodnius prolixus* JH plays an essential role in oogenesis. This hormone makes possible the uptake of vitellogenin by the oocytes (Davey, 1978). It was demonstrated that incubation of follicular cells *in vitro* in the presence of JH leads to a considerable enlargement of the intercellular spaces, so that proteins can penetrate to the oocyte surface, where they are absorbed by pinocytosis. Bell and Barth (1971) demonstrated, by transplanting ovaries to male cockroaches (*Eublaberus posticus*) and injecting the insects with vitellogenin, that the protein was taken up by the oocytes only when JH was given simultaneously with transplantation of ovaries. After administration of JH the ovaries began to accumulate vitellogenins and increase in size, although they did not reach full maturity. An additional condition for the absorption of vitellogenin by the oocytes seems to be the presence of magnesium and calcium ions in the haemolymph (Wajc *et al.*, 1977).

The ability of females to produce vitellogenins is a problem that has attracted the attention of several investigators. Kambysellis (1977) ascertained that the transplantation of *Drosophila melanogaster* ovaries into males of the same species does not inhibit egg maturation. It was also found that the presence of gonads stimulates *de novo* vitellogenin synthesis. The question therefore arose as to how the implanted ovaries stimulate protein synthesis, which does not normally occur in males, and how is it that in various species removal of the ovaries of females does not abolish vitellogenin synthesis in the fat body (Engelmann, 1979). This author also demonstrated that in the case of *Leucophaea maderae* males we are dealing with a different situation; transplantation of ovaries into males interrupts vitellogenesis even when vitellogenin is injected into their haemolymph. Closer observation proved that this substance is actively removed from the haemolymph of the insects. The same situation occurred even when males were implanted with active *corpora allata* together with ovaries, or after administration of JH. We still do not know how this protein, characteristic of the haemolymph of females, is removed from the male body. Similar results have been obtained in males of other species, including *Dictyoptera*, *Orthoptera*, *Hemiptera* and *Coleoptera.* It seems, therefore, that it is not only the males of the *Lepidoptera* that are capable of producing vitellogenins, indicating that the other insect groups possess genetic information for the synthesis of these proteins.

5.4.2.1. Mechanism of action of JH in vitellogenesis

Sufficient evidence has been obtained for the involvement of JH in egg maturation in a number of insect species. The mechanism of this process is not precisely known, however (Wyatt and Pan, 1978). In insects in which JH participates in vitellogenesis, it also stimulates vitellogenin synthesis. Knowing this fact we cannot, however, say that we know the mechanism of action of JH. It can be assumed that JH influences the transcription of DNA into mRNA, but it may also affect ribosomal RNA. This is suggested by the results of experiments with *Leucophaea maderae.* Marked proliferation of ergastoplasm is observed in the cells of the fat body of females during vitellogenesis (Engelmann, 1979). The mechanism of this proliferation, however, is not known, but it is no doubt affected by JH. It is obvious that the more membraneous structures are present within the cytoplasm, the greater the number of ribosomes that can associate with them, and thus, the greater the amount of vitellogenin that can be produced. It is possible that secretion of vitellogenins into the haemolymph by the fat body is also controlled by JH, as suggested by the results of Engelmann and Ladduwahetty

(1974). JH also seems to regulate the absorption of vitellogenin by oocytes (Fig. 5.7).

As seen in this figure, JH may control many stages in the process of egg maturation. Research in the near future will no doubt concentrate on these problems in order to finally elucidate the molecular processes involved in the action of JH.

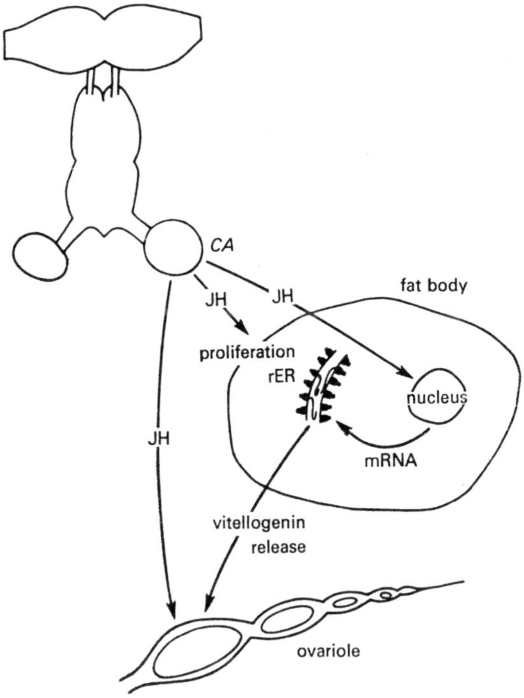

Figure 5.7. Role of JH in vitellogenesis and oogenesis. After Engelmann (1979), with permission.

5.4.3. Role of ecdysone in vitellogenesis

The involvement of ecdysone in reproduction was demonstrated for the first time in studies on the mosquito *Aedes aegypti* (Hagedorn and Kunkel, 1979). After a blood meal ecdysone is released from the ovaries into the haemolymph, and is there transformed to 20-hydroxyecdysone, which stimulates the fat body to synthesize and release vitellogenins into the haemolymph. For these processes to occur, the fat body must have previously been subjected to the action of JH. The fat body probably first reaches a certain state of maturity, as already mentioned in the case of *Locusta migratoria.* Hagedorn *et al.* (1979) obtained important *in vitro* evidence that stimulation of the ovaries to produce ecdysone requires the mediation of the brain hormone. Incubation of ovaries in the presence of brains greatly stimulates synthesis and release of ecdysone by the ovaries. Brains of adult insects have the strongest stimulating ability, whereas brains of larvae are less potent.

In *Orthoptera* and *Lepidoptera* also the ovaries produce ecdysone, but this occurs only towards the end of vitellogenesis, a short time before formation of the chorion (Lagueux *et al.*, 1977; Bollenbacher *et al.*, 1978; Hsiao and Hsiao, 1979). In *Locusta migratoria* ecdysone is produced by the follicular cells, and is released directly into the eggs. In *Lepidoptera* the situation is the same; in these insects ecdysone is also present in the haemolymph, and its function is not known.

Intensive research involving genetic methods is under way, to clarify the mechanisms of action of ecdysone and juvenile hormone in egg maturation (Postlethwait and Handler, 1978; Hagedorn and Kunkel, 1979).

5.4.4. Interaction of hormones during egg maturation

The foregoing review indicates that the production and maturation of eggs in insects depend on many internal and external factors. Food intake is necessary for the stimulation of the *corpora allata* to produce JH, which acts as a gonadotropic hormone. In some species, however, at least one batch of eggs may mature in the absence of feeding. Copulation also stimulates the *CA* to produce JH. The light conditions (photoperiod) also play some role in this process. Thus it seems that, as in the case of metamorphosis, environmental factors synchronize the activity of the endocrine system in the processes of reproduction by the intermediary of the brain, which initiates a chain of reactions stimulating the activity of this system. The interaction of various hormones regulating reproductive processes is relatively well known in mosquitoes (Fig. 5.8). Immediately after the imaginal moult, JH is released and causes oocyte growth until the resting stage is reached. After the female has fed on blood, the neurosecretory cells of the *pars intercerebralis* of the brain release the hormone that stimulates oocyte development (EDNH). This hormone in turn stimulates the ovaries to produce and release ecdysone which, after transformation to 20-hydroxyecdysone, stimulates the fat body to synthesize and release vitellogenins into the haemolymph. The sensitivity of the fat body to ecdysone depends, however, on its previous exposure to juvenile hormone (Flanagan and Hagedorn, 1977). Jowett and Postlethwait (1980) found similarly that the presence of both JH and 20-hydroxyecdysone is necessary for vitellogenin synthesis in the isolated female abdomen of *Drosophila melanogaster*. The mechanism of action of these hormones in the regulation of vitellogenin synthesis in this species is not clear.

However, it seems certain that vitellogenins, under the action of JH, are taken up by the oocytes to form the yolk. Mature eggs are then surrounded by material secreted by the accessory sex glands, the function of which is probably

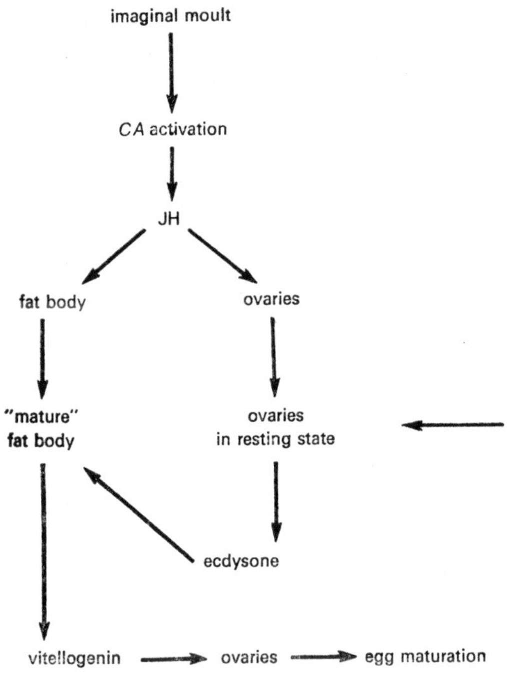

Figure 5.8. Hormonal control of oogenesis in the mosquito. After Riddiford and Truman (1978), with permission.

also regulated by JH (Highnam and Hill, 1977). These glands are usually very small in newly emerged insects, and gradually increase in size with growth of the oocytes. After removal of the *corpora allata* these glands, like the oocytes, do not develop further. It is suspected that the growing oocytes produce a factor stimulating the development of the accessory sex glands. Numerous experiments have shown, however, that the ovaries do not produce such a factor and that the glands are influenced directly by JH (Highnam and Hill, 1977).

5.5. Regulation of the reproductive processes in males

The influence of JH on spermatogenesis has not been demonstrated in male insects. Meiotic division and sperm maturation usually occur during metamorphosis under the influence of ecdysone, and in the absence of JH. In some species meiotic division may start as early as the last larval instar, when the JH level in the haemolymph begins to decrease (see Chapter 4). Schmidt and Williams (1953), however, found that in the haemolymph of *Hyalophora cynthia* males there is a factor essential for the regulation of meiotic division and spermatogenesis in the germarium. These authors demonstrated that spermatocytes from the testes of diapausing pupae do not undergo differentiation in the absence of a substance they termed the macromolecular factor (MF). It appeared that this factor is not species-specific; it is dialysable and thermolabile. These results were confirmed 20 years later by Kambysellis and Williams (1971). It was further ascertained that MF interacts with ecdysone in the regulation of spermatocyte maturation; ecdysone renders the testes permeable to MF, which stimulates spermatocyte maturation in the germarium.

Although JH is not required during spermatocyte maturation, it is indispensable for normal development of the accessory sex glands in the males of some species (Gillot and Friedel, 1976). The secretion of these glands contributes to the formation of the spermatophore and spermatocyte activation, and also produces a change in the reproductive behaviour of females after copulation. In other species, however, the role of JH in these processes has not been definitively confirmed (Tobe *et al.*, 1979).

5.6. Summary

Processes connected with insect reproduction are undoubtedly regulated by hormones. There is much evidence that reproductive behaviour is controlled by juvenile hormone. This hormone also influences the production and excretion of pheromones, a process which conditions certain types of behaviour associated with reproduction.

One of the most interesting aspects of the hormonal regulation of reproduction is the control of vitellogenesis, which has attracted the attention of numerous investigators in recent years. We have begun to see how juvenile hormone controls the synthesis of vitellogenin, and studies in this sphere have progressed to more general problems in endocrinology and developmental biology. Attempts are now being made to discover how JH controls *de novo* protein synthesis by the fully differentiated tissue of the fat body of the adult insect. It is probable that the mechanism of vitellogenin synthesis in insects is similar to that observed in amphibians and birds, making studies on insects still more attractive. In all these animal groups vitellogenins are produced outside the gonads, and insects differ from other groups only in the nature of the hormone controlling vitellogenin production. In insects it is juvenile hormone, whereas in vertebrates it is the steroid hormones. The role of the fat body is fulfilled in vertebrates by the liver. The analogy is even closer if we consider the fact that in males the liver is also capable of vitellogenin production under the influence of oestrogens. In the males of some insect species the fat body can also produce vitellogenins if ovaries are implanted into the abdominal cavity and JH is administered.

The considerable recent developments in research methods will no doubt lead to further discoveries about the hormonal regulation of insect reproduction.

5.7. References

Adams T. S., Hintz A. M., Pomonis J. G. 1968. Oöstatic hormone production in houseflies, *Musca domestica*, with developing ovaries. *J. Insect Physiol.*, **14**; 983–993.

Bollenbacher W. E., Zvenko H., Kumaran A. K., Gilbert L. I. 1978. Changes in ecdysone content during postembryonic development of the wax moth, *Galleria mellonella*. The role of the ovary. *Gen. Comp. Endocrinol.*, **34**; 169–179.

Bell W. J., Barth R. H. 1971. Evidence for juvenile hormone initiated yolk deposition in a cockroach. *Nature*, **230**; 220–221.

Büning J. 1979. The telotrophic-meroistic ovary of *Megaloptera*. I. The ontogenetic development. *J. Morph.*, **162**; 37–66.

Chen T. T., Couble P., Abu-Hakima R., Wyatt G. R. 1979. Juvenile hormone-controlled vitellogenin synthesis in *Locusta migratoria* fat body. Hormonal induction *in vivo*. *Dev. Biol.*, **69**; 70–72.

Couble P., Chen T. T., Wyatt G. R. 1979. Juvenile hormone-controlled vitellogenin synthesis in *Locusta migratoria* fat body. Cytological development. *J. Insect Physiol.*, **25**; 327–337.

Davey K. G. 1978. Hormonal stimulation and inhibition in the ovary of an insect, *Rhodnius prolixus*. [In:] *Comparative Endocrinology* P. J. Gaillard and H. H. Boer (eds.), pp. 13–20. Elsevier, North-Holland, Amsterdam.

Dutkowski A., Cymborowski B. 1970. The diurnal changes in RNA synthesis in the epithelium cells of the egg vesicles of the house cricket. *Bull. Acad. Pol. Sci.*, **17**; 241–244.

Dutkowski A., Cymborowski B. 1971. Role of neurosecretory cells of *pars intercerebralis* in regulating RNA synthesis in some tissues of *Acheta domesticus*. *J. Insect Physiol.*, **17**; 99–108.

Dutkowski A., Cymborowski B. 1973. The role of the suboesophageal ganglion in regulation of RNA synthesis in some insect tissues. *J. Insect Physiol.*, **19**; 1533–1540.

Enders E. 1956. Die hormonale Steuerung rhythmischer Bewegungen von Insekten-Ovidukten. *Verhandl. Deut. Zool. Ges. Erlangen.* **1955**; 113–116.

Engelmann F. 1964. Inhibition of egg maturation in a pregnant viviparous cockroach. *Nature*, **202**; 724–725.

Engelmann F. 1965. The mode of regulation of the *corpus allatum* in adult insects. *Arch. Anat. Microscop. Morphol. Exp.*, **54**; 387–404.

Engelmann F. 1970. *The Physiology of Reproduction.* Pergamon Press, Oxford.

Engelmann F., Ladduwahetty M. 1974. The female specific protein, an extraovarian protein essential for egg maturation in insects. *Zool. Jahrb. Physiol.*, **78**; 289–300.

Engelmann F. 1979. Insect vitellogenin. Identification, biosynthesis and role in vitellogenesis. *Adv. Insect Physiol.*, **14**; 49–108.

Flanagan T. R., Hagedorn H. H. 1977. Vitellogenin synthesis in the mosquito, the role of juvenile hormone in the development of responsiveness to ecdysone. *Physiol. Entomol.*, **2**; 173–178.

Gilbert L. J., Bollenbacher W. E., Granger N. A. 1980. Insect endocrinology: regulation of endocrine glands, hormone titer, and hormone metabolism. *Ann. Rev. Physiol.*, **42**; 493–510.

Gillott G., Friedel T. 1976. Development of accessory reproductive glands and its control by the *corpus allatum* in adult male *Melanoplus sanguinipes. J. Insect Physiol.*, **22**; 365–372.

Hagedorn H. H., Kunkel J. G. 1979. Vitellogenin and vitellin in insects. *Ann. Rev. Entomol.*, **24**; 475–505.

Hagedorn H. H., Shapiro J. P., Hanooka K. 1979. Ovarian ecdysone secretion is controlled by a brain hormone in an adult mosquito. *Nature*, **282**; 92–94.

Highnam K. C., Hill L. 1977. *The Comparative Endocrinology of the Invertebrates*, pp. 131–176. William Clowes and Sons, Ltd., London, Beccles and Clochester.

Hsiao T. H., Hsiao C. 1979. Ecdysteroids in the ovary and the egg of the greater wax moth. *J. Insect Physiol.*, **25**; 45–52.

Jowett T., Postlethwait J. H. 1980. The regulation of yolk polypeptide synthesis in *Drosophila* ovaries and fat body by 20-hydroxyecdysterone and a juvenile hormone analog. *Develop. Biol.*, **80**; 225–234.

Kambysellis M. P. 1977. Genetic and hormonal regulation of vitellogenesis in *Drosophila. Amer. Zool.*, **17**; 535–549.

Kambysellis M. P., Williams C. M. 1971. *In vitro* studies of insects. I. A macromolecular factor prerequisite for silkworm spermatogenesis. *Biol. Bull.*, **141**; 527–540.

Kloc M., Matuszewski B. 1977. Extrachromosomal DNA and the origin of oocytes in the telotrophic-meroistic ovary of *Creophilus maxillosus* (L.) (*Staphylinidae, Coleoptera-Polyphaga*). *Wilhelm Roux's Arch.*, **183**; 351–368.

Kopeć S. 1912. Untersuchungen über Kastration und Transplantation bei Schmetterlingen. *Arch. Entwicklungsmech. Organ.*, **33**; 1–116.

Lagueux M., Hirn M., Hoffmann J. A. 1977. Ecdysone during ovarian development in *Locusta migratoria. J. Insect Physiol.*, **23**; 109–119.

Lazaroviçi P., Pener M. P. 1978. The relations of the *pars intercerebralis*, *corpora allata*, and juvenile hormone to oocyte development and oviposition in the African migratory locust. *Gen. Comp. Endocrinol.*, **35**; 375–386.

Lea A. O. 1963. Some relationship between environment, *corpora allata*, and egg maturation in *Aedine mosquitoes. J. Insect Physiol.*, **9**; 793–809.

Lea A. O. 1967. The medial neurosecretory cells and egg maturation in mosquitoes. *J. Insect Physiol.*, **13**; 419–429.

Leopold R. A. 1976. The role of male accessory glands in insect reproduction. *Ann. Rev. Entomol.*, **21**; 199–221.

Martoja R. 1964. Un type particulier d'appareil génital féminin chez les insectes: les ovarioles adénomorphes du Coléoptère *Steraspis speciosa* (*Heterogastra*, *Buprestidae*). *Bull. Soc. Zool. Fr.*, **89**; 614–641.

Mesnier M. 1980. Study of oocyte growth within the ovariole in a stick-insect, *Clitumnus extradentatus*. *J. Insect Physiol.*, **26**; 59–65.

Naisse J. 1965. Contrôle endocrinien de la différenciation sexuelle chez les insectes. *Arch. d'anat micr. exper.*, **54**; 417–428.

Nayar K. K. Studies on the neurosecretory system of *Iphita limbata* Stal. V. Probable endocrine basis of oviposition in the female insect. *Proc. Indian Acad. Sci.*, **47**; 233–251.

Postlethwait J. H., Handler A. M. 1978. Non-vitellogenic female sterile mutants and the regulation of vitellogenesis in *Drosophila melanogaster*. *Dev. Biol.*, **67**; 202–213.

Riddiford L. M., Truman J. W. 1978. Biochemistry of insect hormones and insect growth regulators. [In:] Biochemistry of Insect, pp. 307–357. Academic Press. New York, San Francisco, London.

Riddiford L. M. 1980. Insect endocrinology: Action of hormones at the cellular level. Ann. Rev. Physiol., **42**; 511–528.

Schmidt E. L., Williams C. M. 1953. Physiology of insect diapause. V. Assay of the growth and differentiation hormone of *Lepidoptera* by the method of tissue culture. *Biol. Bull.*, **105**; 174–187.

Strangways-Dixon J. 1961. The relationship between nutrition, hormones and reproduction in the blowfly *Calliphora erythrocephala* Meig. I. Selective feeding in relation to the reproductive cycle, the *corpus allatum* volume and fertilization. *J. Exp. Biol.*, **38**; 225–235.

Telfer W. H. 1954. Immunological studies of insect metamorphosis. II. The role of a sex-limited blood protein in egg formation by the *Cecropia* silkworm. *J. Gen. Physiol.*, **37**; 539–558.

Telfer W. H. 1975. Development and physiology of the oocyte nurse cell syncytium. *Adv. Insect Physiol.*, **11**; 223–319.

Thomas A. 1979a. Nervous control of egg progression into the common oviduct and genital chamber of the stick-insect *Carausius morosus*. *J. Insect Physiol.*, **25**; 811–823.

Thomas A. 1979b. Inhibition of the onset of vitellogenesis by abnormal retention of mature eggs in the stick insect, *Carausius morosus*. *Intern. J. Invert. Reprod.*, **1**; 347–354.

Thomsen E. 1948. Effect of removal of neurosecretory cells in the brain of adult *Calliphora erythrocephala* Meig. *Nature*, **161**; 439–440.

Tobe S. S., Musters A., Stay B. 1979, *Corpus allatum* function during sexual maturation of male *Diploptera punctata*. *Physiol. Entomol.*, **4**; 79–86.

Truman J. W., Riddiford L. M. 1977. Invertebrate systems for the study of hormonal effects on behavior. *Vitam. Horm. N.Y.*, **35**; 283–315.

Wajc E., Baker-Grunwald T., Applebaum S. W. 1977. Binding and uptake of trypan blue by developing oocytes of *Locusta migratoria migratorioides*. *J. Embryol. exp. Morphol.*, **37**; 1–11.

Wigglesworth V. B. 1936. The function of the *corpus allatum* in the growth and reproduction of *Rhodnius prolixus* (*Hemiptera*) *Quart. J. Microscop. Sci.*, **79**; 91–121.

Wilde J., de, De Loof A. 1973. Reproduction—endocrine control. [In:] *The Physiology of Insecta*. M. Rockstein (ed.), vol. 1; pp. 91–157. Academic Press, New York.

Wilde J., de 1964. Reproduction—endocrine control. [In:] *Physiology of Insecta*, M. Rockstein (ed.), vol. 1; pp. 59–90. Academic Press, New York.

Wilkens J. L. 1968. The endocrine and nutritional control of egg maturation in the fleshfly *Sarcophaga bullata*. *J. Insect Physiol.*, **14**; 927–943.

Wyatt G. R., Pan M. L. 1978. Insect plasma proteins. *Ann. Rev. Biochem.*, **47**; 779–817.

Chapter 6

Insect diapause

6.1. General characteristics of diapause

Insects have evolved special mechanisms allowing them to survive in highly diverse environments. One such adaptation is diapause—a periodical interruption of development, making survival possible under unfavourable environmental conditions. Insects in the state of diapause have a very low metabolic rate and show special biochemical adaptations, such as the increased glycerol concentration in the haemolymph of saturnid pupae which allows them to tolerate very low winter temperatures. Before entering diapause insects usually build up large reserves of fat, protein and carbohydrate. They become resistant to dehydration by producing a special wax coating, which is particularly important in tropical species.

Knowing the characteristic changes accompanying true diapause, it can easily be distinguished from pseudodiapause caused, for instance, by low or high temperature, dry air or lack of food. Another characteristic of physiological diapause is that it can be interrupted only when the accompanying processes are ended. For instance, interruption of pupal diapause in *Hyalophora cecropia* can occur only after 6 weeks of chilling (at about 3°C) and after a further two weeks at an ambient temperature of 25°C (Williams, 1946). Thus, these unfavourable conditions during diapause are necessary for the reactivation of development.

6.1.1. Types of diapause

In insects having a long postembryonic period, lasting one year or more, diapause occurs regularly at a particular stage of development, notwithstanding the prevailing environmental conditions. Such diapause is termed obligatory, and usually occurs in species with a univoltine life cycle, In multivoltine species diapause occurs only in the generation that develops during periods of unfavourable environmental conditions. An example of such a species is the large white butterfly (*Pieris brassicae*). From eggs laid in early spring hatch larvae that complete development into adults. These lay eggs in late summer, and the larvae that emerge develop into pupae. These pupae diapause over the winter to produce the next generation in spring, and so on. This type of diapause is termed facultative, because it depends on environmental conditions. That this is so may be checked by breeding the insect in the laboratory under conditions characteristic of summer (long days and high temperature). Under such conditions this species does not enter diapause.

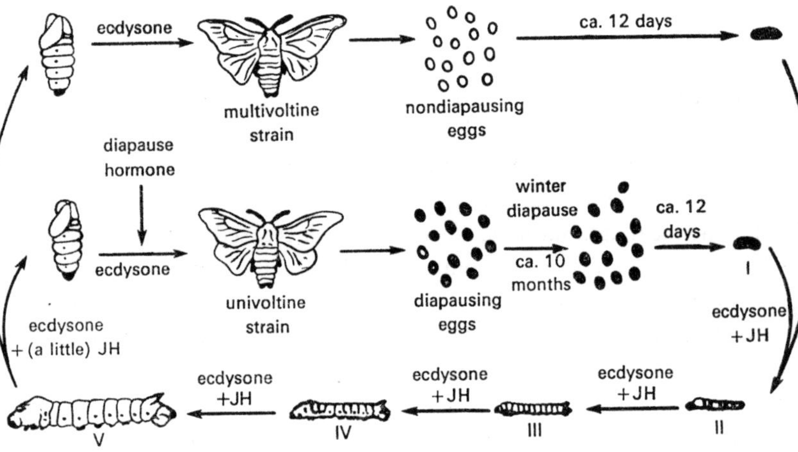

Figure 6.1. Life cycle of the silkworm *Bombyx mori*. Univoltine and multivoltine strains of this species are represented. Further details in text. After Isobe and Goto (1980), with permission.

There are two strains of the silkworm (*Bombyx mori*): a multivoltine and a univoltine one (Isobe and Goto, 1980). The multivoltine strain lays nondiapausing eggs, so that many generations of adult insects develop in a single year (Fig. 6.1). The univoltine strain lays eggs which undergo diapause, so that only one generation appears during the summer.

6.2. Environmental factors and diapause

Insect diapause may be evoked by environmental factors such as temperature, moisture and the quality and availability of food. These factors change with the seasons, but since they show wide variations within any given environment and are not good indicators of seasonal changes, it is difficult to forecast the sequence of seasons from them.

The duration of daylight (the photoperiod; the number of hours of illumination in each 24-h period) is the factor that changes in a predictable and precise way during the course of the year. It is perhaps for this reason that insects "chose" the photoperiod as the decisive factor in the induction of diapause.

6.2.1. Influence of photoperiod on induction of diapause

After metamorphosis, the great majority of insects, reproduce in summer, so that imposition of a short photoperiod leads them to enter diapause (Fig. 6.2).

Numerous experiments have demonstrated that insects are highly sensitive to a critical photoperiod, below which they enter diapause. In some species a change as small as 10–15 min in 24 h causes a considerable change in the percentage of diapausing insects. As seen in Figure 6.2, a difference of about one hour is usually sufficient to make all the insects enter diapause or to continue development (Saunders, 1976).

The curve relating insect responses to photoperiod length is usually plotted on the basis of experimental data on insects kept under a constant photoperiod, longer or shorter than the critical value. Such conditions of course never prevail in the natural environment. The natural photoperiod undergoes continuous

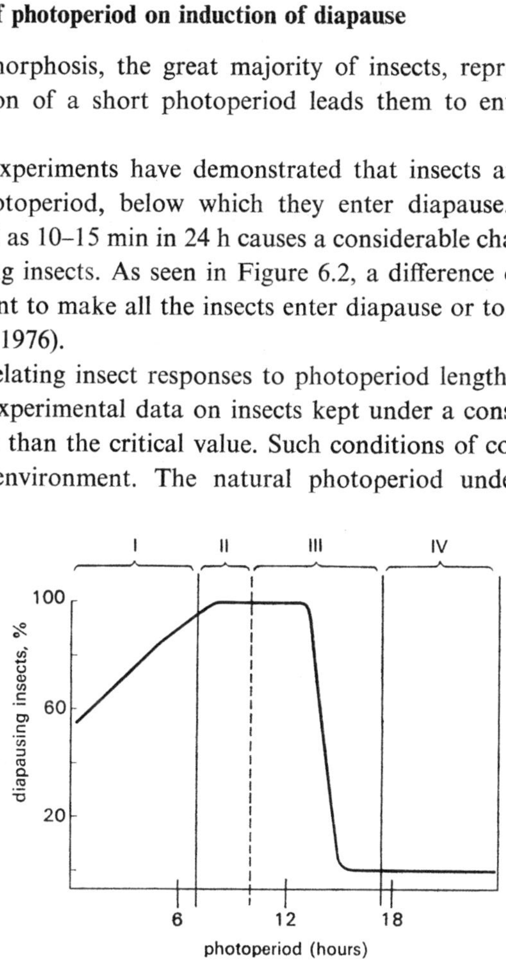

Figure 6.2. Influence of photoperiod on insect diapause. Vertical continuous lines show range of variation of natural photoperiod. Compartments I and IV represent photoperiods that do not occur under natural conditions. Compartment II represents photoperiodic conditions in winter, when the temperature is below the minimum necessary for insect development. Only compartment III is ecologically important. After Saunders (1976), with permission.

variation: it increases or decreases continuously. The question thus arises as to the sensitivity of insects to increases or decreases in the photoperiod.

It seems, however, that such a problem does not exist for insects, since they are not sensitive to photoperiod length throughout their entire period of development. In some species, for instance, the sensitive period is in autumn, and in others it is in spring. In *Bombyx mori* the most sensitive stage is the egg, although the first larval stage is also somewhat sensitive. In *Philosamia cynthia* only larval stages 4 and 5 are sensitive to the short photoperiod that induces pupal diapause. In the development of *Diataraxia oleracea* the period of sensitivity lasts only two days, and occurs at the end of the penultimate instar (Highnam and Hill, 1977). There is thus something like a delayed reaction to the photoperiod, since the period of sensitivity precedes the developmental stage that undergoes diapause by a considerable interval.

According to de Wilde (1962), the photoperiod may act in three different ways:

1. photoperiodic induction of diapause which is a reversible process,
2. photoperiodic excitation of the centres controlling growth and development,
3. a photoperiodic reaction of the stimulated systems.

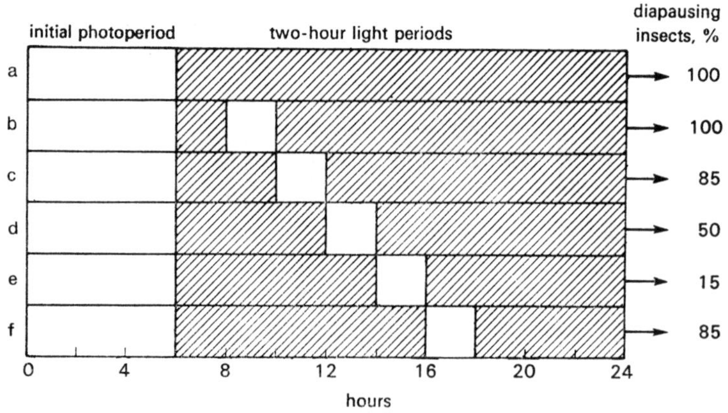

Figure 6.3. Influence of 2-hour periods of light at various times during the night on pupal diapause in *Pieris brassicae* reared under a cycle of 6 h light and 18 h darkness, in which 100 per cent of the insects enter diapause (a). When the 2-h light period starts at noon, the percentage of diapausing pupae diminishes to 50 (d). Diapause occurs in only a few insects after exposure to light between 2 p.m. and 4 p.m. (e). Illumination at other hours has little or no effect on diapause (b, c, f). After Highnam and Hill (1977), with permission.

Further investigations showed that not only does sensitivity to the photoperiod appear at a particular developmental stage, but also that the extent of the response to light differs even within a 24-h period in some species (Beck, 1980).

For instance, if *Pieris brassicae* larvae are reared under a light regime of 6 h daylight and 18 h darkness, 100 per cent of the pupae developing from them enter diapause (Fig. 6.3). When the period of darkness is interrupted by a two-hour period of illumination at various times during the 24-h period, the percentage of diapausing insects is considerably less; the insects react as if they were under long-day conditions. This experiment suggests the existence of diurnal changes in the sensitivity of the larvae to light.

6.2.2. Photoperiod and developmental reactivation

Studies by many authors have shown that insects entering diapause in autumn when the daylength falls below the critical value resume development in spring, when the length of the photoperiod is increasing. In these insects the reactivation of development also depends on the photoperiod. Williams and Adkisson (1964) demonstrated that the photoperiod that induces pupal diapause in *Antheraea pernyi* is an exact reflection of the photoperiod necessary for developmental reactivation (Fig. 6.4). This suggests that the biological clock that measures

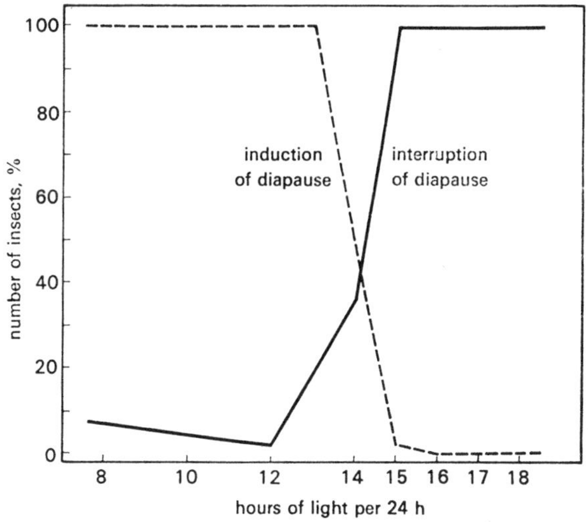

Figure 6.4. Photoperiodic induction and interruption of pupal diapause in *Antheraea pernyi*. Note that the curve of the critical photoperiod length inducing diapause is a mirror image of that of the photoperiod interrupting diapause. After Williams and Adkisson (1964).

the length of the photoperiod takes part in the regulation of both of these processes (Cymborowski, 1976).

Other environmental factors such as temperature and nutrition also participate in the reactivation of development, as we will see in further parts of this chapter.

6.2.3. Mechanism of action of the photoperiod

Numerous studies on the influence of the photoperiod on diapause indicate that in insects there is usually one period of sensitivity to light during each 24-hour period, and sometimes two periods of sensitivity. These periods occur at different times in the diurnal cycle in different species (Saunders, 1976). The period of sensitivity plays an essential role in the induction and interruption of diapause.

What is the mechanism of photoactivation? It is postulated that during the period of sensitivity to light substances are produced in the body which stimulate the insect to further development (Fig. 6.5A). If the development of the insect occurs over a sufficiently long period, the time of sensitivity to light coincides with the period of illumination that led to the production of the above-mentioned

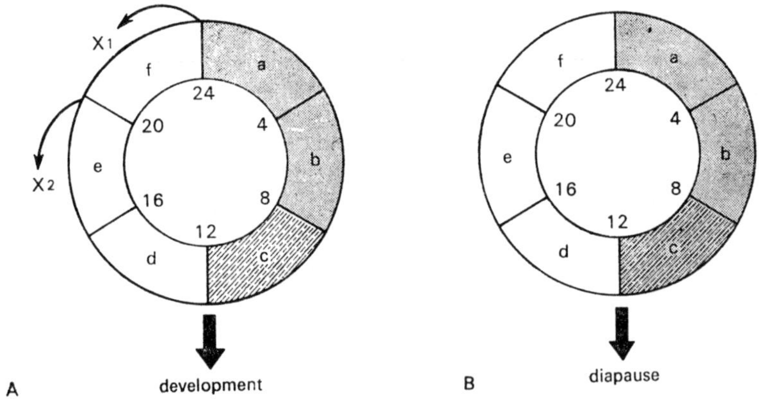

Figure 6.5. Theoretical model of the effect of photoperiod on insect diapause. Several reactions (biochemical processes) occur consecutively (a, b, f) resulting in production of factor(s) (x_1, x_2) essential for the control of development or the prevention of diapause. The whole cycle lasts about 24 h and at least one of its stages is activated by light. In this diagram it is stage C (shaded area). In the long photoperiod (A) this stage occurs during the light period, so that activation occurs and factors preventing diapause are produced. Under the short photoperiod (B) the light-sensitive phase occurs during the dark period (dark and shaded areas); the factor x (x_1, x_2) is not produced and the insect enters diapause.

substances. If the insect is exposed to a short photoperiod, however (Fig. 6.5 B), photoinduction does not occur and diapause takes place.

Photoinduction requires a certain number of light cycles. We are dealing here, therefore, with the phenomenon of a cumulative effect of photoperiod (Danilevskii, 1965; Saunders, 1976; Beck, 1980). It may thus be assumed that information on the photoperiodic conditions accumulates in the brain (see Chapter 4), and that the mechanism responsible for this consists of at least two elements: the biological clock, which allows the insect to distinguish long days from short days, and a counter which sums the successive cycles until the moment when diapause is induced or development is reactivated.

Different insect species require particular constant numbers of light cycles in order that photoinduction can occur. Photoinduction is affected by temperature, and it would seem that some substance accumulates during exposure to a long photoperiod (probably some neurosecretion from the brain). Its release stimulates development of the insect. Exposure to a short photoperiod does not allow accumulation of the necessary amount of this neurosecretion, so that development is inhibited and diapause occurs.

It is possible that this mechanism is more complicated, and it may be that under a long photoperiod a factor is produced in the brain which releases neurosecretory material which would lead to continuation of development. Under a short photoperiod, this factor is not produced, and although neurosecretion is accumulated it is not released into the haemolymph, and consequently the insect diapauses. Another possible explanation is that during the short photoperiod a substance is produced in the brain which inhibits the release of neurosecretory material. These problems will no doubt be solved in the near future.

6.2.4. Influence of temperature on diapause

Under natural conditions the photoperiod is not the only factor that determines diapause. Other factors may be temperature and food. Temperature exerts its influence in various ways. Constant temperature often modifies the degree of response to a particular photoperiod. It may, for instance, change the length of the photoperiod that induces diapause.

Under natural conditions, high temperatures generally accompany long days, and both promote development. Low temperatures and a short photoperiod, on the contrary, induce diapause. In insects such as *Pieris brassicae*, temperature up to about 26°C has little influence on the critical photoperiod length (Danilevskii, 1965). In the aphid *Megoura viciae* the critical photoperiod length decreases by about 15 min with each 5°C increase in temperature, but above 23°C the insect ceases to react to short days (Lees, 1964).

In other insect species temperature may completely abolish the influence of a particular photoperiod. Figure 6.6 shows the modifying action of temperature in *Sarcophaga argyrostoma* (Saunders, 1971). Under a photoperiod of 10 h light and 14 h darkness and a temperature range of 15–18°C, almost 100 per cent of pupae enter diapause. On the other hand, under the same photoperiod but with a temperature range of 20–25°C, diapause does not occur. High temperature may thus completely abolish the effect of short photoperiod.

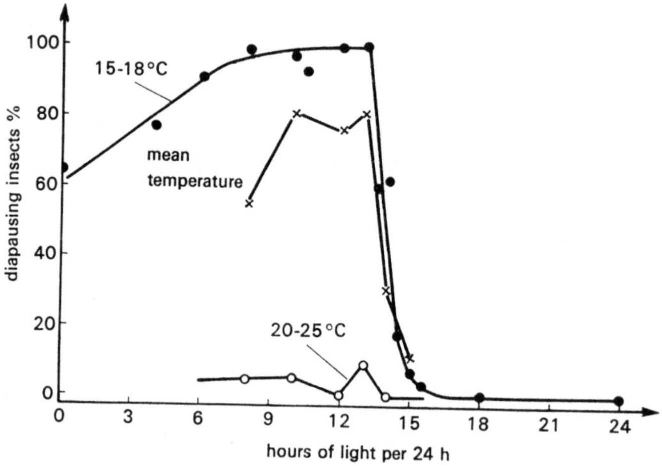

Figure 6.6. Influence of temperature on photoperiodic reaction in *Sarcophaga argyrostoma*. Description in text. After Saunders (1971).

This phenomenon is common (Cullen and Browning, 1976), although cases are known in which high temperature stimulates diapause, as in *Abraxas miranda* (Masaki, 1958). This is understandable, since this species normally develops under conditions of short photoperiod and low temperature.

It may be said in general that there is an optimal temperature range, similar to that occurring in nature, which interacts with the photoperiodic conditions in the induction of diapause (Danilevskii, 1965). Species living in a cold climate require low temperatures. In *Saturnia pavonia*, for instance, the optimal temperature for the induction of diapause lies within the range −7°C to +7°C.

For tropical insects the optimal temperatures for diapause may be very high, reaching 27–38°C in the case of *Oedipoda miniata* (Orshan and Pener, 1979a, b).

6.2.5. Influence of nutrition on diapause

The food of insects is never the same over the course of the year because its quality and quantity change with the seasons. For instance, herbivorous insects such as the larvae of *Lepidoptera* have young leaves at their disposal in spring, but in autumn there are only yellow leaves of greatly reduced nutrient value. Predaceous insects must also adapt themselves to the diminishing numbers of prey.

As in the case of temperature, a particular diet may modify the response to photoperiod. In some insects this factor plays an essential role in the cycle of their seasonal activity. Adkisson (1961) and Akdisson *et al.* (1963) demonstrated that, depending on the quality and quantity of oil added to the diet of *Pectinophora gossipiella* larvae, a short photoperiod (10 h light and 14 h darkness) causes 15–80 per cent of the larvae to diapause. Under natural conditions larvae of these insects feed on cotton seeds. It is probable that the oil content of these seeds increases as they ripen, stimulating the insects to diapause. We are dealing here with a response to a short photoperiod, distinctly modified by the diet.

Bradshaw (1970) similarly demonstrated that interruption of larval diapause in *Chaoborus americanus* requires long days and the availability of food in the form of mosquito larvae. Transfer of the diapausing larvae to the laboratory and provision of a large quantity of food causes about 50 per cent of the insects to interrupt their diapause in spite of exposure to a short photoperiod. When the transferred insects are starved but subjected to a long photoperiod, about 50 per cent of the larvae undergo further development, but after being fed under these conditions nearly all of them break their diapause. This is evidence that food is acting synergistically with the long photoperiod in the developmental reactivation of the diapausing insects.

This brief review indicates that food plays an essential role both in the induction of diapause (interacting with a short photoperiod) and in the reactivation of development (interacting with a long photoperiod).

6.3. Hormonal regulation of diapause

6.3.1. Embryonic diapause

In many insect species diapause occurs at some stage during embryonic development in the egg. Usually it takes place during blastokinesis, more or less in the middle of embryonic development, but it may also occur towards the end of development, just before hatching, in completely formed larvae. Although many cases of embryonic diapause are known, the hormonal regulation of this process has been studied in detail only in the silkworm (*Bombyx mori*). The presence of a diapause hormone has also been detected in *Orygia antigua* (*Limantriidae*).

6.3.1.1. The diapause hormone

Bombyx mori females lay eggs that do or do not undergo diapause, depending on the photoperiodic conditions in which they live. When female larvae are reared under short photoperiods, such as prevail in spring, the adult insects that develop from them produce eggs that do not undergo diapause, whereas adults derived from larvae reared under a long photoperiod produce diapausing eggs. Hasegawa (1951) found that in the latter case the suboesophageal ganglion of the female secretes a special diapause hormone that inhibits development of the eggs laid by acting on the ovaries. Detailed investigations showed that this hormone is released from the neurosecretory cells of the suboesophageal ganglion in the pupal period (Hasegawa, 1952). Fukuda (1962) obtained similar results. He also showed that secretion of the diapause hormone by the suboesophageal ganglion is controlled by the brain via a nervous pathway.

Soon afterwards the biologically active hormone was isolated from the suboesophageal ganglion (Hasegawa, 1957) and was studied intensively. It appears to play an important role in the metabolic processes associated with diapause. For instance, it acts on the pupal ovaries, increasing their permeability to 3-hydroxykinurenine, the precursor of the pigment that appears in diapausing eggs (Ogawa and Hasegawa, 1975). It also stimulates the activity of trehalase in the pupal ovaries, leading to synthesis and increased accumulation of glycogen in the diapausing eggs compared with nondiapausing eggs (Yamashita and Hasegawa, 1976). It also blocks activity of esterase A, which is involved in the mobilization of the yolk proteins necessary for embryonic development (Kai and Hasegawa, 1973; Kai and Nishi, 1976). Egg reactivation requires a long period of chilling, during which esterase A is activated in an unknown manner. When the eggs are transferred to a higher temperature, esterase A induces lysis of the yolk-containing cells, allowing embryonic development to resume.

Isobe and Goto (1980) carried out intensive investigations on the chemical structure of the diapause hormone and its physico-chemical properties. Earlier Kubota *et al.* (1976) reported that the diapause hormone consists of two active fractions, A and B, with molecular weight 3300 and 2000 respectively. After purification both these fractions become highly sensitive to temperature and light

(Kubota *et al.*, 1979). Chemical analyses demonstrated that both forms of this hormone are lipoproteins. Hormone A is less active than B (Fig. 6.7). The differences in the activity of the two forms may be due to differences in their rates of inactivation; however, the mechanisms are so far unknown.

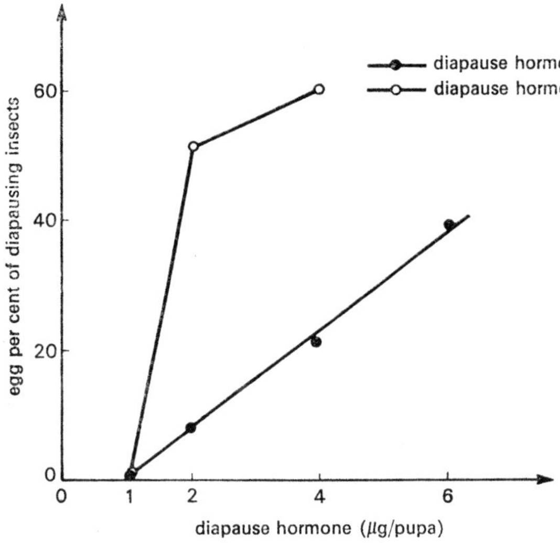

Figure 6.7. Biological activity of diapause hormones A and B. Activity was tested on 4-day-old pupae of *Bombyx mori*. Modified according to Isobe and Goto (1980), with permission.

6.3.2. Larval diapause

Larval diapause takes place in the last instar as a rule, although it may set in at earlier larval stages. In *Choristoneura fumiferana*, for instance, diapause occurs in the second larval stage (Harvey, 1961). *Dendrolimus pini* enters diapause in different larval stages depending on the climatic conditions prevailing in the given year (Geyspitz, 1965).

6.3.2.1. Characteristics of larval diapause

Larval diapause is characterized by a greatly depressed metabolic rate, a reduction (in terrestrial insects) of body water content, conspicuous accumulation of fat reserves, and inhibition of growth and metamorphosis. In spite of reduced motor activity, diapausing larvae retain the capacity for coordinated movement. The larvae of *Chilo suppressalis* and *Diatraea grandiosella* undergo facultative diapause towards the end of the last instar. In both species the characteristic features are cessation of feeding, a low rate of respiratory metabolism, and inhibition of spermatogenesis. However, permanent larval moults continue (Fukaya and Mitsuhashi, 1961; Chippendale and Yin, 1973). Similar signs of larval diapause are observed in *Ostrinia nubilalis*, but without continuous larval moults (Chippendale and Yin, 1979).

In the great majority of insects larval diapause is interrupted by exposure to a particular photoperiod. Temperature also plays an important role in the reactivation of development (Chippendale, 1977).

6.3.2.2. Role of JH in regulation of larval diapause

At the beginning of the 1960s (Beck, 1964; Beck and Alexander, 1964) it was suggested that the larval diapause of *Ostrinia nubilalis* is controlled by a hormone produced by the gut. This hormone was named proctolin. Investigations by other authors, however, did not confirm these results, but drew attention to the role of juvenile hormone in the regulation of larval diapause. The first evidence for the participation of JH in this process was obtained in studies on *Chilo suppressalis* larvae (Fukaya and Mistuhashi, 1961) and larvae of *Diatraea grandiosella* (Chippendale and Yin, 1973). Yagi and Akaike (1976) presented proof that both induction and maintenance of diapause in *Ostrinia nubilalis* depend on a high JH concentration in the haemolymph. Yin and Chippendale (1979) also demonstrated that the activity of the *corpora allata* is necessary during larval diapause in *Diatraea grandiosella*, because removal of these glands causes premature interruption of diapause, whereas treating the larvae with a JH analogue after removal of the *CA* prevents this interruption. These authors also demonstrated that the brain maintains *CA* activity via a nervous pathway. Measurement of the JH concentration in the haemolymph of diapausing and nondiapausing insects demonstrated important differences, especially during the last larval instar (Fig. 6.8). While in nondiapausing larvae (Fig. 6.8 A) a drastic fall in the JH level occurs as early as the first day after the moult to the last instar, in diapausing larvae the hormone level rises (Fig. 6.8 B). The authors also demonstrated that administration of a JH analogue to nondiapausing larvae of the last instar causes them to enter into a diapause with features similar to those of physiological diapause.

The data given here indicate that larval diapause does not occur as a result of depression of the activity of the endocrine system; on the contrary, the system is active during diapause. This is true at least for the *corpora allata*. The brain

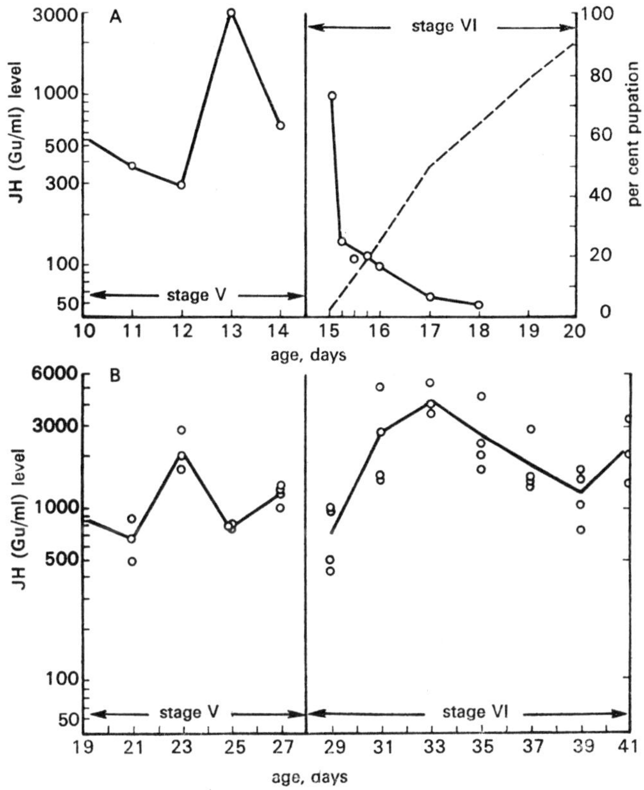

Figure 6.8. Juvenile hormone level in penultimate and last larval stage of nondiapausing (A) and diapausing (B) *Diatraea grandiosella*. Broken line shows percentage of pupation. The JH concentration was determined by the *Galleria* test. After Yin and Chippendale (1979), with permission.

and prothoracic glands must also function if permanent larval moults occur frequently during larval diapause.

6.3.3. Pupal diapause

6.3.3.1. Characteristics of pupal diapause

A pupal diapause is the most common state in which insects pass the winter. It occurs mainly in *Lepidoptera* and *Diptera*, but also in other insect groups. Its most characteristic feature is a considerable depression of metabolism, manifested in a low respiratory rate associated with a reduced number of functioning mitochondria in tissues other than muscle. All processes associated with imaginal differentiation are interrupted. The diapausing pupae are highly resistant to dehydration and low temperature, owing to the large quantity of glycerol present in their bodies.

Pupal diapause is induced by short photoperiods during the larval period. Reactivation of development, on the contrary, occurs only after several weeks or even months of exposure to low temperature (usually below 10°C). The pupae are not usually sensitive to the photoperiod, although in *Hyalophora cecropia* and *Antheraea pernyi* they are still sensitive to light conditions after entering diapause. Hence, diapause can be prolonged by exposure to a short photoperiod, whereas it is interrupted when the pupae are transferred for several weeks to a long photoperiod (Shakhbazov, 1961). These results were confirmed by Williams and Adkisson (1964). It was also noted that light acts directly on the brain of the pupa, changing its secretory activity (Williams *et al.*, 1965).

6.3.3.2. Hormonal control of pupal diapause

Hormonal regulation of pupal diapause is well known through the discoveries of Williams (1946, 1947). He found in *Platysamia cecropia* that it is the result of inhibition of prothoracicotropic activity of the brain. Production and release of PTTH are inhibited by short photoperiods, so that ecdysone, necessary for normal development, is not secreted (see Chapter 4).

The mechanism of action of short and long photoperiods on the secretory activity of the insect brain is not yet known. Kono (1973, 1977) found wide differences in the histological structure of type *A* neurosecretory cells in the *pars intercerebralis* of the brain of diapausing and nondiapausing pupae of *Pieris rapae*. In diapausing individuals cytoplasmic secretory granules are more abundant than in nondiapausing pupae.

The question therefore arises as to the cause of the inhibition of PTTH release in diapausing pupae. Van der Kloot (1955, 1961) suggests that it is caused by the absence of bioelectric activity in the brain of diapausing pupae. This does not seem to be true for the larvae of *Bupalus piniarius* (Schoonhoven, 1962), in which the brain exhibits spontaneous bioelectric activity accompanied by neurosecretion. Exposure to chilling is necessary for the release of the neurosecretory material.

The rate of PTTH release during chilling of diapausing pupae differs among insect species. In *Hyalophora cecropia* it is a very slow process. Complete release of this hormone occurs as late as the end of diapause. On the other hand, in diapausing pupae of *Mimas tiliae* (*Sphingidae*) the process is much quicker. Highnam (1958) demonstrated that in the first three weeks of exposure of these pupae to 3°C, the whole of the neurosecretory material that has accumulated passes to the *corpora cardiaca* and is rapidly released, causing interruption of diapause. Activation of the neurosecretory system during chilling of diapausing pupae of *Pieris rapae* was confirmed by the results of electron microscope studies (Kono, 1977).

Relatively little is known about the mechanism of activation of the endocrine system of diapausing pupae by temperature, which, as in larval diapause, plays an important role in the reactivation of development. Rasenick *et al.* (1976) demonstrated that transfer of pupae of *Antheraea pernyi* from a temperature of 5°C to room temperature (ca. 25°C) increases as much as six-fold the cAMP content in the brain compared with that in diapausing pupae. It was found in further investigations that about 80 per cent of the cAMP detected is present in the *pars intercerebralis*, i.e. where the neurosecretory cells are localized (Rasenick *et al.*, 1978). It would seem, therefore, that activation of the brain in diapausing pupae may be initiated by an increase in cAMP level in the neurosecretory cells, which would stimulate further PTTH release. A similar mechanism seems to function in the reactivation of development of diapausing *Antheraea pernyi* pupae after exposure to a long photoperiod (Rasenick *et al.* 1978).

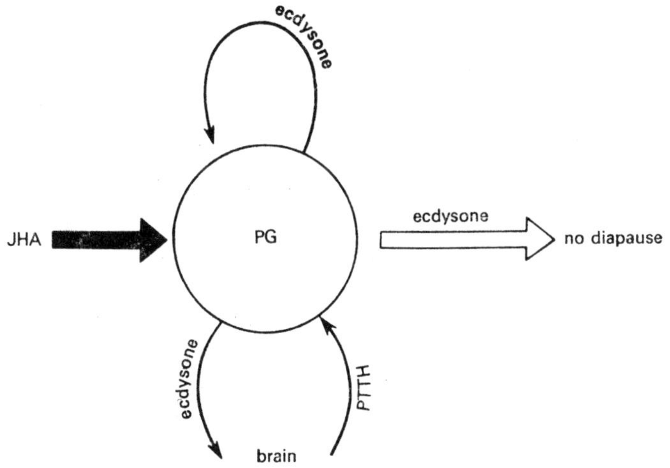

Figure 6.9. Probable mechanism of action of juvenile hormone analogue (JHA) during interruption of pupal diapause in *Mamestra brassicae*. After Hiruma (1979), with permission. PG—prothoracic glands, PTTH—prothoracic hormone. Further explanation in text.

Recently attention has been called to the possibility that juvenile hormone is involved in the interruption of diapause. Hiruma (1979) demonstrated that administration of a JH analogue (ZR-515) to *Mamestra brassicae* larvae in the last stage of development prevents pupal diapause. Similar results were obtained with *Manduca sexta* larvae (Bradfield and Denlinger, 1980). To prevent pupal diapause, however, the insects had to be injected with a large dose of JH analogue (ZR-515).

The mechanism of action of JH in this process is not yet known in detail. Studies on *Manduca sexta* showed that this hormone may act directly on the prothoracic glands of mature larvae of the last instar and on pupae, stimulating ecdysone production (Safranek *et al.*, 1980). The ecdysone released may itself act on the prothoracic glands, stimulating them to further ecdysone production by positive feedback. This hormone may also stimulate the production and release of PTTH (Fig. 6.9).

Further studies will probably show which of these elements plays the essential role in the prevention of diapause by JH. The fact that *Manduca sexta* pupae are capable of initiating development even when the brain is removed (Judy, 1972; Cymborowski, 1981) seems to indicate that the brain is not an essential link in the stimulation of ecdysone synthesis and release.

6.3.4. Adult diapause

6.3.4.1. Characteristics of adult diapause

Diapause in adult insects is characterized above all by inhibition of the reproductive function. Diapausing insects show many characteristic changes in behaviour and metabolism. Processes such as feeding, motor activity, and reactions to various light and mechanical stimuli are also modified. For instance, newly moulted Colorado beetles (*Leptinotarsa decemlineata*) show positive phototaxis. After entering diapause they cease to feed, and begin to display negative phototaxis; they descend from the host plant and bury themselves in the soil (de Wilde and de Boer, 1961; de Wilde, 1962). Many other insects also leave their host plant and search for a suitable shelter for diapause. Such behaviour is observed in *Hypera postica* and *Anthonomus grandis*, and in many ladybird species (Hodek and Tcherkasov, 1963).

The rate of oxygen uptake in diapausing adult insects is usually lower than that in nondiapausing insects. Large fat and glycogen reserves accumulate in diapausing insects (Tombes, 1966). This is particularly important since diapausing insects do not feed. During diapause gonad development also ceases, especially in females. In many beetles, flies and mosquitoes the ovaries remain undeveloped until feeding is resumed (see Chapter 5).

6.3.4.2. Hormonal regulation of adult diapause

Hormonal regulation of imaginal diapause is best known from the studies of de Wilde (1963, 1965) on the Colorado beetle (*Leptinotarsa decemlineata*). This author found that removal of the *corpora allata* from nondiapausing beetles leads to a state characteristic of normal diapause. Implantation of active *CA* into these insects interrupts diapause.

Thus, under conditions of short photoperiod the production and release of JH are inhibited, probably owing to the inhibitory action of the brain on the *CA*. This leads to imaginal diapause. This hypothesis has been confirmed by many authors (de Wilde and de Boer, 1969; Schooneveld, 1970; Hodková, 1977).

The levels of various juvenile hormones have been measured in diapausing and nondiapausing *Leptinotarsa decemlineata* (Deleurance *et al.*, 1980). It was found that in pupae from which nondiapausing insects later eclosed the JH II level was much higher than in pupae from which diapausing insects eclosed (Fig. 6.10). On the other hand, in beetles entering diapause the JH III content was much lower than that in nondiapausing individuals. No difference was, however, found in ecdysteroid concentrations, suggesting that these hormones are not directly involved in the regulation of imaginal diapause.

6.4. Summary

In order to survive unfavourable environmental conditions many insect species enter diapause, characterized by inhibition of metabolic processes and adaptive biochemical changes. Owing to their ability to "measure" the photoperiod by means of the biological clock, insects are able to "foresee" the occurrence of unfavourable conditions and to prepare in advance for diapause.

Apart from the photoperiod and temperature the quantity and quality of food play an essential role in the induction of diapause.

The phenomenon of diapause may occur at practically any stage of insect development in the embryo, larva, pupa and adult (Fig. 6.11). Embryonic diapause is induced by a long photoperiod leading to secretion of the diapause hormone by the female suboesophageal ganglion, inhibiting embryonic development in

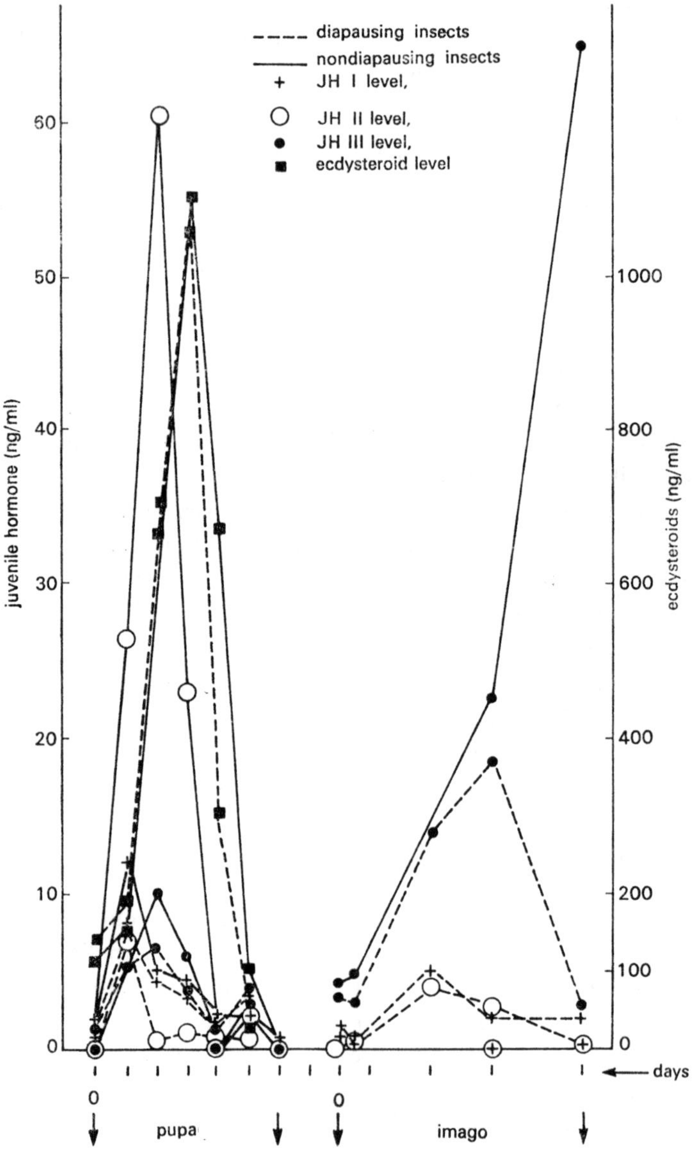

Figure 6.10. Juvenile hormone and ecdysterone concentrations in haemolymph of nondiapausing (continuous line) and diapausing (broken line) pupae and imagines of *Leptinotarsa decemlineata*. Note high JH III level in nondiapausing beetles compared with diapausing ones. After Deleurance (1980), with permission.

the eggs that are laid. Larval diapause is produced by a short photoperiod, which induces a high JH level in the body, and this prevents metamorphosis. Other endocrine centres (brain, *corpora cardiaca* and prothoracic glands) seem to be active during larval diapause, as indicated by the frequently observed permanent larval moults. Pupal diapause is characterized by inhibition of the prothoracicotropic activity of the brain on exposure to short photoperiod. This leads to inhibition of ecdysone production and release. Reactivation of development occurs mainly through the action of long photoperiods, but ambient temperature also plays a role here. Imaginal diapause takes place under short photoperiods, but in this case the allatotropic activity of the brain is inhibited, leading to a considerable depression of the JH level in the body; this makes gonad development impossible. Thus, imaginal diapause is above all associated with inhibition of reproductive processes in insects.

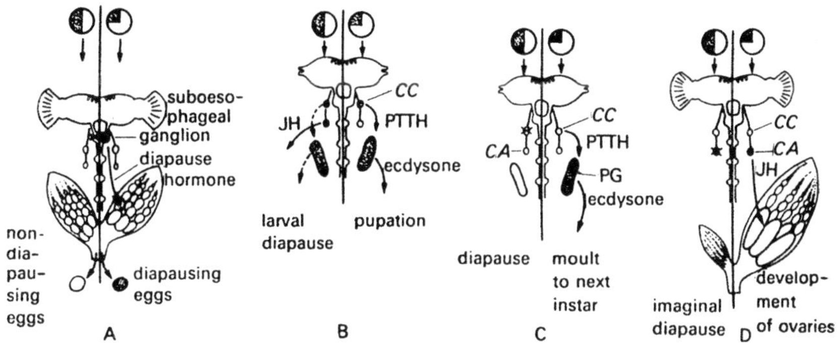

Figure 6.11. Four types of insect diapause. After Saunders (1976), with permission. A—embryonic, B—larval, C—pupal, D—imaginal. Active glands are dotted, inactive ones are crossed out. *CC*—*corpora cardiaca*, PG—prothoracic glands, PTTH—prothoracicotropic hormone. Hatched parts of circles denote periods of darkness during the 24-h cycle.

6.5. References

Adkisson P. L. 1961. Effect of larval diet on the seasonal occurrence of diapause in the pink bollworm. *J. Econ. Entomol.*, **54**; 1107–1112.

Adkisson P. L., Bell R. A., Wellso S. G. 1963. Environmental factors controlling the induction of diapause in the pink bollworm, *Pectinophora gossipiella* (Saunders). *J. Insect Physiol.*, **9**; 299–310.

Beck S. D. 1964. Time measurement in insect photoperiodism. *Amer. Natural.*, **98**; 329–346.

Beck S. D., Alexander N. 1964. Chemically and photoperiodically induced diapause development in the European corn borer, *Ostrinia* nubilalis. *Biol. Bull.*, **126**; 175–184.

Beck S. D. 1980. *Insect Photoperiodism*. Second Edition. Academic Press, New York, London, Toronto, Sydney, San Francisco.

Bowers W. S., Blickenstaff C. C. 1966. Hormonal termination of diapause in the alfalfa weevil. *Science*, **154**; 1673–1674.

Bradshaw W. E. 1970. Interaction of food and photoperiod in the termination of larval diapause in *Chaoborus americanus* (*Diptera Culicidae*). *Biol. Bull.*, **139**; 476–484.

Bradfield J. Y., Denlinger D. L. 1980. Diapause development in the tobacco hornworm: A role for ecdysone or juvenile hormone? *Gen. Comp. Endocrinol.*, **41**; 101–107.

Chippendale G. M., Yin C. M. 1973. Endocrine activity retained in diapause insect larvae. *Nature*, **246**; 511–513.

Chippendale G. M. 1977. Hormonal regulation of larval diapause. *Ann. Rev. Entomol.*, **22**; 121–138.

Chippendale G. M., Yin C. M. 1979. Larval diapause of the European corn borer, *Ostrinia nubilalis*: Further experiments examining its hormonal control. *J. Insect Physiol.*, **25**; 53–58.

Cullen J. M., Browning T. O. 1976. The influence of photoperiod and temperature on the induction of diapause in pupae of *Heliothis punctigera*. *J. Insect Physiol.*, **24**; 595–601.

Cymborowski B. 1976. *Zegary biologiczne* [*Biological clocks*]. PWN, Warszawa.

Cymborowski B. 1981. On the necessity and unnecessity of the brain in insect development. [In:] *Regulation of Insect Development and Behaviour*. F. Sehnal, A. Zabża, J. J. Menn, B. Cymborowski (eds.), pp. 21–36. Technical University Press, Wrocław.

Danilevskii A. S. 1965. *Photoperiodism and Seasonal Development of Insects*. Oliver and Boyd, Edinburgh and London.

Deleurance S., Baehr J-C., Porcherron P., Gassier P. 1980. Diapause et évolution des taux des ecdysteroides et des hormones juvéniles chez les nymphes et les imagos de *Leptinotarsa decemlineata* Say. *C.R. Acad. Sci.*, **290**; 367–370.

Fukaya M., Mitsuhashi J. 1961. Larval diapause in the rice stem borer with special reference to its hormonal mechanism. *Bull. Nat. Inst. Agric. Sci.* (*Japan*), **13C**; 1–32.

Fukuda S. 1962. Hormonal control of diapause in the silkworm. *Gen. comp. Endocrinol. Suppl.*, **1**; 337–340.

Geyspitz, K. F. 1965. Photoperiodic and temperature reactions affecting the seasonal development of the pine moths, *Dendrolimus pini* L. and *D. sibiricus* Tschetw. *Entomol. Rev.*, **44**; 316–325.

Harvey G. T. 1961. Second diapause in spruce budworm from eastern Canada. *Can. Entomol.* **93**; 594–602.

Hasegawa K. 1951. Studies on the voltinism in the silkworm, *Bombyx mori* L. with special reference to the organs concerning determination of voltinism. *Proc. Japan Acad.*, **27**; 667–671.

Hasegawa K. 1952. Studies of the voltinism in the silkworm *Bombyx mori* L. with special reference to the organs concerning determination of voltinism. *J. Fac. Agric. Tattori. Univ.*, **1**; 83–126.

Hasegawa K. 1957. The diapause hormone of the silkworm *Bombyx mori*. *Nature*, **179**; 1300–1301.

Highnam K. C. 1958. Activity of the brain *corpora cardiaca* system during pupal diapause "break" in *Mimas tiliae* (*Lepidoptera*). *Quart. J. micr. Sci.*, **99**: 73–88.

Highnam K. C., Hill L. 1977. Endocrine mechanisms in the *Insecta*. III. Other endocrine controlled processes. [In:] *The Comparative Endocrinology of the Invertebrates*, pp. 177–192. William Clowes and Sons Ltd., London, Beecles and Clochester.

Hiruma K. 1979. Prevention of pupal diapause by the application of juvenile hormone analogue to the last instar larvae of *Mamestra brassicae*. *Appl. Entomol. Zool.*, **14**; 76–82.

Hodek J., Čerkasov J. 1963. Imaginal dormancy in *Semiadalia undecimnolata* Schneid. II. Changes in water, fat and glycogen content. *Acta Soc. Zool. Bohem.*, **27**; 298–318.

Hodkova M. 1977. Function of the neuroendocrine complex in diapausing *Pyrrhocoris apterus* females. *J. Insect Physiol.*, **23**; 23–28.

Isobe M., Goto T. 1980. Diapause hormones. [In:] *Neurohormonal Techniques in Insects*. T. A. Miller (ed.), pp. 216–243. Springer-Verlag, New York, Heidelberg, Berlin.

Judy K. J. 1972. Diapause termination and metamorphosis in brainless tobacco hornworm (*Lepidoptera*). *Life Sci.*, **11**; 605–611.

Kai H., Hasegawa K. 1973. An esterase in relation to yolk cell lysis at diapause termination in the silkworm, *Bombyx mori*. *J. Insect Physiol.*, **19**; 799–810.

Kai H., Nishi K. 1976. Diapause development in *Bombyx* eggs in relation to "esterase A" activity. *J. Insect Physiol.*, **22**; 1315–1320.

Kono Y. 1973. Light and electron microscopic studies on the neurosecretory control of diapause incidence in *Pieris rapae crucivora*. *J. Insect Physiol.*, **19**; 255–272.

Kono Y. 1977. Ultrastructural changes of neurosecretory cells in the *pars intercerebralis* during diapause development in *Pieris rapae*. *J. Insect Physiol.*, **23**; 1461–1473.

Kubota I., Isobe M., Goto T., Hasegawa K. 1976. Molecular size of the diapause hormone of the silkworm *Bombyx mori*. *Z. Naturforsch.*, **31c**; 132–134.

Kubota I., Isobe M., Imai K., Goto T., Yamashita O., Hasegawa K. 1979. Characterization of the silkworm diapause hormone B. *Agric. Biol. Chem.* (*Japan*). **43**; 1075–1078.

Lees A. D. 1964. The role of photoperiod and temperature in the determination of parthenogenetic and sexual forms in the aphid *Megoura viciae* Buckton. III. Further properties of the maternal switching mechanism in apterous aphids. *J. Insect Physiol.*, **9**; 153–164.

Masaki S. 1958. The response of a "short-day" insect to certain external factors: the induction of diapause in *Abraxas miranda Bul. Japan J. appl. Entomol. Zool.*, **2**; 28–294.

Ogawa H., Hasegawa K. 1975. Kynurenine-3-hydroxylase activity and follicle development in the silkworm, *Bombyx mori*. *Insect Biochem.*, **5**; 119–134.

Orshan L., Pener M. P. 1979a. Termination and reinduction of reproductive diapause by photoperiod and temperature in males of the grasshopper, *Oedipoda miniata*. *Physiol. Entomol.*, **4**; 55–61.

Orshan L., Pener M. P. 1979b. Repeated reversal of the reproductive diapause by photoperiod and temperature in males of the grasshopper, *Oedipoda miniata*. *Entomol. exp. appl.*, **25**; 219–226.

Rasenick M. M., Neuburg M., Berry S. J. 1976. Brain cyclic AMP levels and the initiation of adult development in the *Cecropia* silkmoth. *J. Insect. Physiol.*, **22**; 1453–1456.

Rasenick M. M., Neuburg M., Berry S. J. 1978. Cyclic nucleotide activation of the silkmoth brain-cellular localization and further observation on the patterns of activation. *J. Insect. Physiol.*, **24**; 137–139.

Safranek L., Cymborowski B., Williams C. M. 1980. Effect of juvenile hormone on ecdysone-dependent development in the tobacco hornworm, *Manduca sexta*. *Biol. Bull.*, **158**; 248–256.

Saunders D. S. 1971. The temperature compensated photoperiodic clock "programming" development and pupal diapause in the fleshfly, *Sarcophaga argyrostoma*. *J. Insect. Physiol.*, **17**; 801–812.

Saunders D. S. 1976. *Insect Clocks*. Pergamon Press, Oxford, New York, Toronto, Sydney, Paris, Frankfurt.

Schoonhoven L. M. 1962. Diapause and the physiology of host parasite synchronization in *Bupalus piniarius* L. (*Geometridae*) and *Eucarcelia rutilla* Vill. (*Tachinidae*). *Arch. Neer. Zool.*, **15**; 111–174.

Schooneveld H. 1970. Structural aspects of neurosecretory and *corpus allatum activity* in the adult Colorado beetle, *Leptinotarsa decemlineata* Say. as a function of daylength. *Netherl. J. Zool.*, **20**; 151–237.

Shakhbazov V. G. 1961. The reaction to the length of daylight and the light receptor for the pupa of the Chinese oak silkworm, *Antheraea pernyi* G. *Dokl. Akad. Nauk. SSSR*, **140**; 944–946.

Tombes A. S. 1966. Aestivation (summer diapause) in *Hypera postica* (*Coleoptera*: *Curculionidae*). I. Effect of aestivation, photoperiods and diet on total fatty acids. *Ann. Entomol. Soc. Am.*, **59**; 376–380.

Wilde J., de, de Boer J. A. 1961. Physiology of diapause in the adult Colorado beetle. II. Diapause as a case of pseudoallatectomy. *J. Insect Physiol.*, **6**; 152–161.

Wilde J., de, 1962. Analysis of the diapause syndrome in the Colorado potato beetle (*Leptinotarsa decemlineata* Say): Behavior and reproduction. *Acta Physiol. Pharmacol. Neerl.*, **11**; 525.

Wilde J., de, 1963. Photoperiodic control of *corpus allatum* activity in the Colorado beetle. *Acta Physiol. Pharmacol. Neerl.*, **12**; 344.

Wilde J., de, 1965. Photoperiodic control of endocrines in insects. *Arch. Anat. Microscop. Morphol. Exp.*, **54**; 547–564.

Wilde J., de, de Boer J. A. 1969. Humoral and nervous pathways in photoperiodic induction of diapause in *Leptinotarsa decemlineata*. *J. Insect Physiol.*, **15**; 661–675.

Williams C. M. 1946. Physiology of insect diapause: the role of the brain in the production and termination of pupal dormancy in the giant silkworm *Platysamia cecropia*. *Biol. Bull.*, **90**; 234–243.

Williams C. M. 1947. Physiology of insect diapause. II. Interaction between the pupal brain prothoracic glands in the metamorphosis of the giant silkworm, *Platysamia cecropia*. *Biol. Bull.*, **90**; 89–98.

Williams C. M., Adkisson P. L. 1964. Physiology of insect diapause. XIV. An endocrine mechanism for the photoperiodic control of pupal diapause in the oak silkworm, *Antheraea pernyi*. *Biol. Bull.*, **127**; 511–524.

Williams C. M., Adkisson P. L., Walcott C. 1965. Physiology of insect diapause. XV. The transmission of photoperiod signals to the brain of the oak silkworm, *Antheraea pernyi*. *Biol. Bull.*, **128**; 479–507.

Van der Kloot W. G. 1955. The control of neurosecretion and diapause by physiological changes in the brain of the *cecropia* silkworm. *Biol. Bull.*, **109**; 276–294.

Van der Kloot W. G. 1961. Insect metamorphosis and its endocrine control. *Amer. Zoologist.*, **1**; 3–9.

Yagi S., Akaike N. 1976. Regulation of larval diapause by juvenile hormone in the European corn borer, *Ostrinia nubilalis*. *J. Insect Physiol.*, **22**; 389–392.

Yamashita O., Hasegawa K. 1976. Diapause hormone action in silkworm ovaries incubated *in vitro*, ^{14}C-trehalose incorporation into glycogen. *J. Insect Physiol.*, **22**; 409–414.

Yin C-M., Chippendale G. M. 1979. Diapause of the southwestern corn borer, *Diatraea grandiosella*: Further evidence showing juvenile hormone to be the regulator. *J. Insect Physiol.*, **25**; 513–523.

Chapter 7

Hormonal regulation of insect metabolism

7.1. Hormonal regulation of respiratory metabolism

Measurement of oxygen uptake by insects under various experimental conditions, including interference with the functions of the endocrine system, has frequently been performed in order to obtain indirect information on the mechanism of action of insect hormones. Artificial lowering or raising of the titre of hormones in investigations of respiratory metabolism allows the examination of the role of these hormones in the energetic processes of the body. A search was made over many years for an insect metabolic hormone similar in function to thyroxine in vertebrates. To date, however, no such hormone has been detected. Nevertheless, much information has been obtained about the effect on respiratory metabolism of hormones regulating growth and metamorphosis, especially juvenile hormone. It was expected that hormones which influence processes leading to complete reconstruction of the body would be involved in its general metabolism. This was confirmed by numerous observations on the changes in oxygen utilization during different developmental stages of insects (Niemierko, 1956). It appeared that the processes associated with larval or imaginal moulting are accompanied by characteristic changes in the rate of oxygen uptake (Sláma, 1965; Sehnal and Sláma, 1966; Sláma *et al.*, 1964; Sláma and Hodková, 1975).

7.1.1. Involvement of JH in respiratory metabolism

Numerous investigations demonstrated that disturbance of the developmental rhythm of insects by implantation of active *corpora allata* (*CA*) or administration of juvenile hormone and its analogues leads to characteristic changes in metabolic processes, and this is reflected in the rate of O_2 uptake (Sehnal and Sláma, 1966; Sehnal *et al.* 1976; Kryspin-Sørensen *et al.*, 1977). Maximal oxygen consumption by developing larvae usually occurs at the middle of instar (Fig. 7.1). Before and during moulting O_2 uptake decreases, whereas in the pupal period there is first a decrease in the rate of uptake, followed by an increase towards the end of the period. The curve of oxygen utilization is therefore U-shaped.

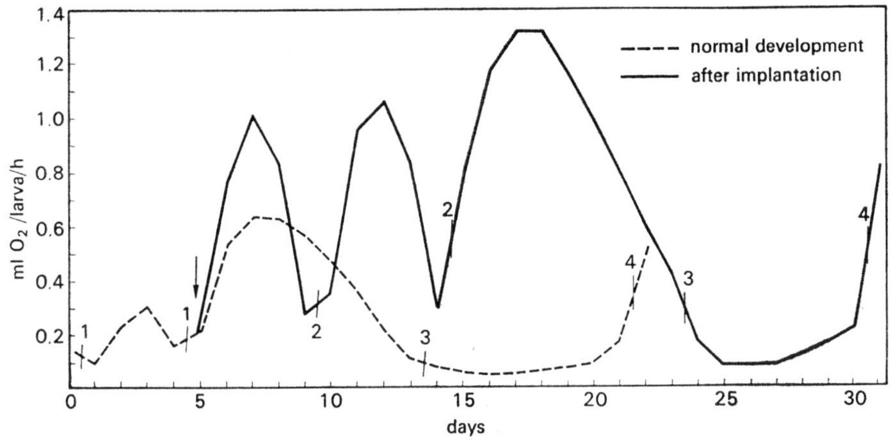

Figure 7.1. Changes in oxygen uptake during normal development of *Galleria mellonella* larvae (broken line) and after implantation of brain + *corpora cardiaca* + *corpora allata* (continuous line). Arrow indicates time of implantation. Vertical lines denote successive moults. After Sehnal and Sláma (1966).
1—larval moult, 2—superlarva, 3—pupal moult, 4—imaginal moult.

Similar changes in respiratory metabolism also occur in the additional larval stages arising after the implantation of active *corpora allata* (see Fig. 7.1). This indicates the significant role of JH in the regulation of the rate of oxygen uptake. The involvement of JH is confirmed by the results of earlier experiments in which removal of the *CA* during the larval stage caused a drastic depression of respiratory metabolism in *Dixippus morosus*. Reimplantation of *CA* restored the characteristic pattern of changes in oxygen utilization (L'Hélais, 1954, 1964; Lüscher and Leuthold, 1965).

Other endocrine centres besides the *corpora allata* also seem to affect insect respiratory metabolism. For instance, removal of the *corpora cardiaca* (*CC*) together with the *corpora allata* in *Pyrrhocoris apterus* decreases oxygen uptake to a greater extent than removal of the *CA* alone (Sláma, 1964). Reimplantation of *CC* into adult *Leucophaea maderae* females from which both the *CC* and the *CA* have been removed increases oxygen uptake by 60 per cent compared with control insects (Müller and Engelmann, 1968). In adult males and females of *Schistocerca gregaria*, removal of the *CC* produces a 30 per cent decrease in oxygen utilization about 10 days after the operation (Lafont-Gazal and Roussel, 1971). Similarly, removal of the medial neurosecretory cells of the brain in *Oncopeltus fasciatus* leads to a considerable decrease in oxygen uptake (Conradi-Larsen, 1970).

The results presented here are, however, difficult to interpret because in none of these studies were JH levels determined in the haemolymph of the insects subjected to removal or implantation of endocrine glands. It is, therefore, quite probable that these treatments caused changes in the titre of JH, leading to changes in respiratory metabolism. Moreover, since food is a very important factor influencing insect respiratory metabolism (Wojtczak, 1952; Kryspin, 1977), it is possible that the operations described could have interfered with the uptake and digestion of food, and this could affect the rate of oxygen uptake.

7.1.2. The phenomenon of hypermetabolism and its mechanism

Investigations on larvae of *Dermestes maculatus* showed that in this species JH and its analogues do not cause morphogenetic changes similar to those usually observed in other insects (Sláma, 1976); however, after treatment with an analogue of JH, larvae show more than 5-fold increase in oxygen uptake (Kryspin, 1977; Sláma and Kryspin-Sørensen, 1979). This phenomenon was called hypermetabolism (Fig. 7.2). Its occurrence, as shown in further investigations, depends on the developmental stage of the insect and on the quality of the food consumed. Hypermetabolism can be evoked only in the penultimate and last larval instars,

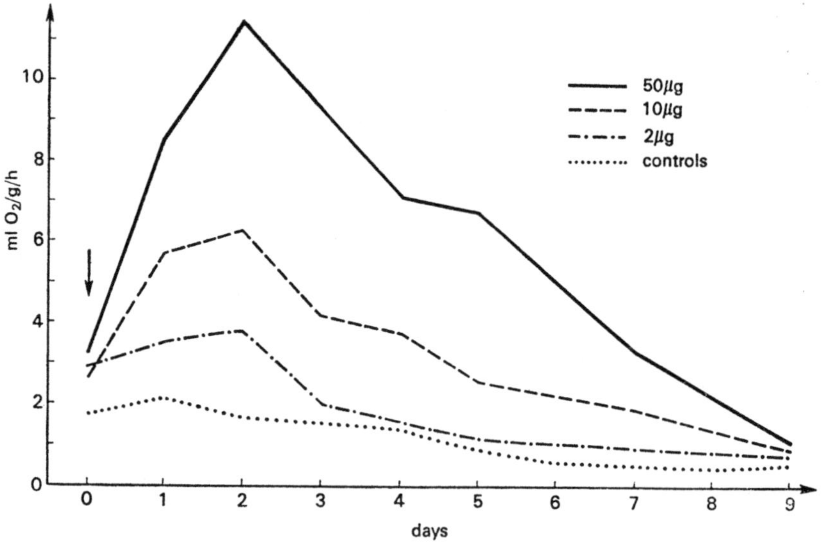

Figure 7.2. Oxygen uptake by final instar *Dermestes maculatus* larvae after administration of various doses of juvenile hormone analogues. Arrow indicates time of hormone administration. After Sláma and Kryspin-Sørensen (1979), with permission.

and when food rich in high-energy components such as fats is available to the insect. Treatment of last-instar larvae with a JH analogue, starving them, or feeding them with delipidated food do not evoke hypermetabolism. When starved larvae are treated with JH simultaneously with access to normal food, oxygen uptake immediately increases (Fig. 7.3).

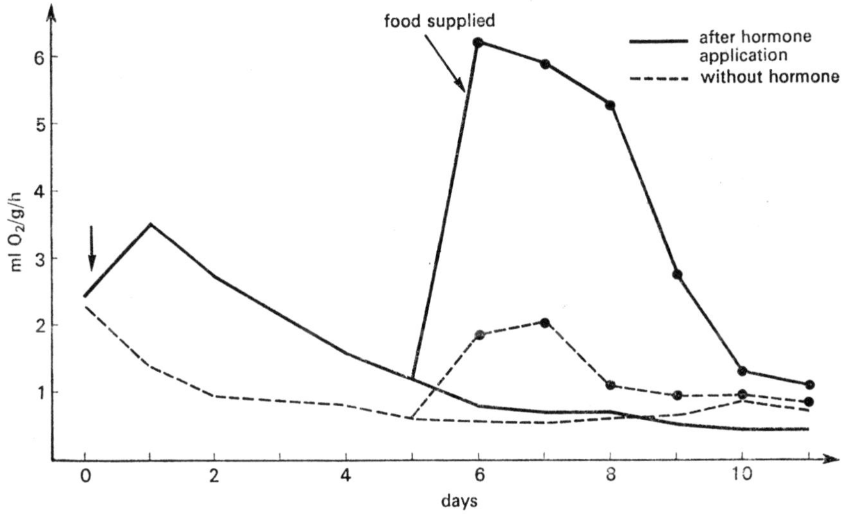

Figure 7.3. Oxygen uptake by final-instar *Dermestes maculatus* larvae after application of 100 μg juvenile hormone analogue (continuous line) and by control larvae (broken line). The larvae were starved for 5 days after hormone application (arrow) and then received food. Further details in text. After Sláma and Kryspin-Sørensen (1979), with permission.

To elucidate the mechanism of hypermetabolic action of JH on *Dermestes vulpinus* larvae, the influence of this hormone on the activity of enzymes of the respiratory chain needs to be studied. Investigations of this type have been conducted for more than 20 years. Clark and Baldwin (1960) showed that addition of *CA* homogenate to the fat body *in vitro* increases the respiratory metabolism of the mitochondria of this organ. Further intensive biochemical studies indicated that the activity of enzymes of the respiratory chain is a very sensitive index of the action of JH and its analogues (Minks, 1967; Keeley, 1970, 1971; Firstenberg and Silhaček, 1973). It appeared that JH inhibits oxidation of NAD-dependent substrates such as pyruvate, whereas malate, α-ketoglutarate and glutamate stimulate the oxidation of succinate (Silhaček *et al.*, 1974). Kryspin (1977) demonstrated in hypermetabolic *Dermestes vulpinus* larvae enhanced activity of two mitochondrial enzymes, NADH oxidoreductase: cytochrome C and succinate dehydrogenase, in the gut, fat body and Malpighian tubules. The changes were most pronounced in the fat body, in which a 30-fold increase in succinate dehydrogenase activity was noted (Fig. 7.4).

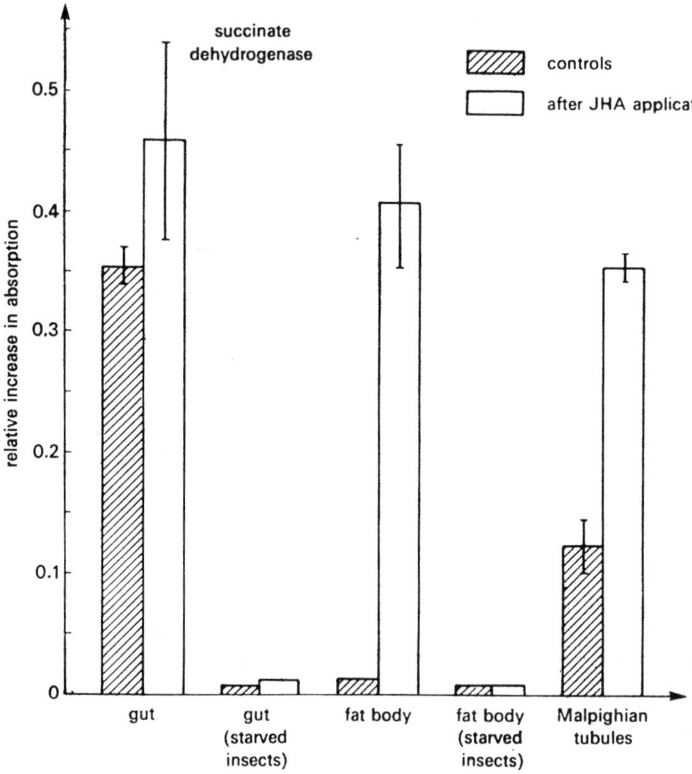

Figure 7.4. Changes in succinate dehydrogenase activity in organs of *Dermestes maculatus* treated with juvenile hormone analogue (JHA) and untreated control insects. Further details in text. After Kryspin (1977), with permission.

Besides these changes in enzyme activity in various tissues, it was found that JH and its analogues may cause uncoupling of oxidative phosphorylation in the mitochondria under *in vivo* conditions (Stegawee and van Kammen-Werthen, 1962). This process involves the inhibition of ATP production—electron transport remaining unaffected—resulting in the production of heat and metabolic water. This is associated with a considerable increase in succinate dehydrogenase

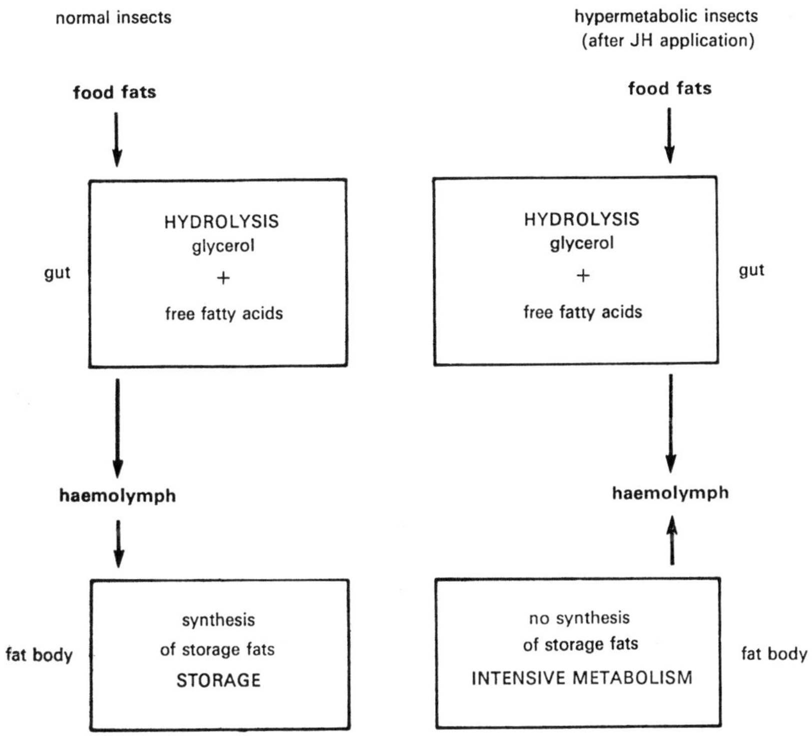

Figure 7.5. Probable mechanism of action of juvenile hormone in hypermetabolism of *Dermestes maculatus* larvae. Application of JH analogue causes release of free fatty acids from the fat body into the haemolymph; these enhance respiratory metabolism. Explanation in text.

activity (Wojtczak, 1969). Hypermetabolism may therefore result from changes in oxidative phosphorylation caused by JH. However, this effect of JH is apparently not direct. It was found that administration of JH or an analogue enhances fat metabolism, leading to accumulation of free fatty acids in the haemolymph (Gilbert, 1967; Sroka and Barth, 1976). Free fatty acids, on the other hand, may cause uncoupling of oxidative phosphorylation in the mitochondria (Wojtczak and Wojtczak, 1960; Wojtczak *et al.*, 1969). In hypermetabolic insects a change in fat metabolism is thus observed under the influence of JH, leading to a high oxygen requirement (Fig. 7.5).

The results presented here concerning the hormonal regulation of insect metabolism allow the following generalizations:

1. removal of *corpora allata* leads to a considerable decrease in oxygen uptake by insects,

2. implantation of *corpora allata*, especially into allatectomized insects, causes a rise in oxygen uptake,

3. *corpora cardiaca* stimulate respiratory metabolism of the fat body,

4. in some insects JH and its analogues evoke hypermetabolism.

7.2. Carbohydrate metabolism

It has long been known that in vertebrates the concentration of carbohydrates, especially glucose, is regulated by hormones. Insulin is a hypoglycaemic hormone. depressing the blood sugar level, and glucagon acts as a hyperglycaemic hormone increasing the sugar concentration. There is now much evidence suggesting that insects use a similar mechanism for the regulation of sugar concentration in the haemolymph. The main carbohydrate in the haemolymph is the disaccharide trehalose, although a small quantity of glucose is also present (Wyatt, 1967).

7.2.1. Hormonal regulation of trehalose concentration in insect haemolymph

7.2.1.1. The hyperglycaemic hormone

The hyperglycaemic factor, similar to the glucagon of vertebrates, was detected in the *corpora cardiaca* of *Periplaneta americana* (Steele, 1961, 1963). Further investigations confirmed the presence of this hormone in various insect groups (Goldsworthy, 1969; Normann and Duve, 1969; Normann, 1975; Tager *et al.*, 1975 and others). The hyperglycaemic hormone is of a proteinaceous nature, with a molecular weight of 4500; the molecule is therefore about 30 times larger than mammalian glucagon. Its activity is abolished by trypsin.

Many authors suggest that the hyperglycaemic hormone is synthesized by the medial neurosecretory cells of the brain and stored in the *corpora cardiaca* (Wiens and Gilbert, 1967). If this is so, the hormone must be released very rapidly, since it is practically undetectable in the *CC* (Steele, 1961). The hormone is possibly synthesized by the *CC* cells in which it is usually detected.

The hyperglycaemic action of this hormone consists in the activation of glycogen phosphorylase in the fat body, probably with the cooperation of cyclic AMP. It has been shown that an extract of the *corpus cardiacum* of *Periplaneta americana* increases the cAMP level in the fat body after as little as five minutes.

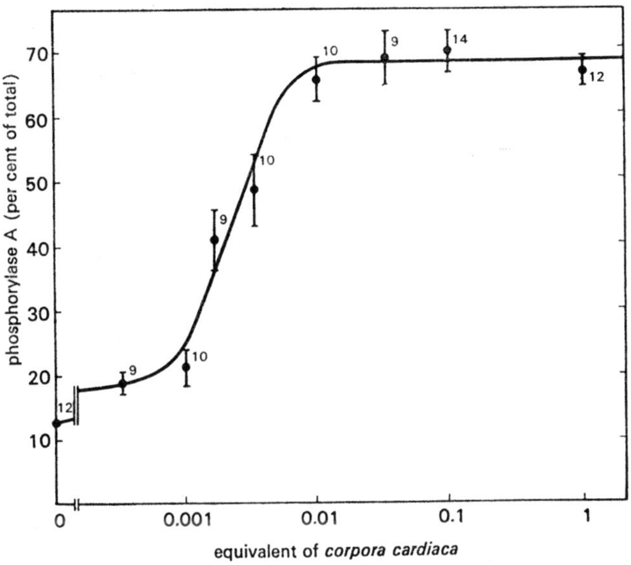

Figure 7.6. Stimulation of glycogen phosphorylase activity of final-instar *Manduca sexta* larvae after application of hyperglycaemic hormone. The figures on the curves denote the number of larvae used for the tests. After Ziegler (1979), with permission.

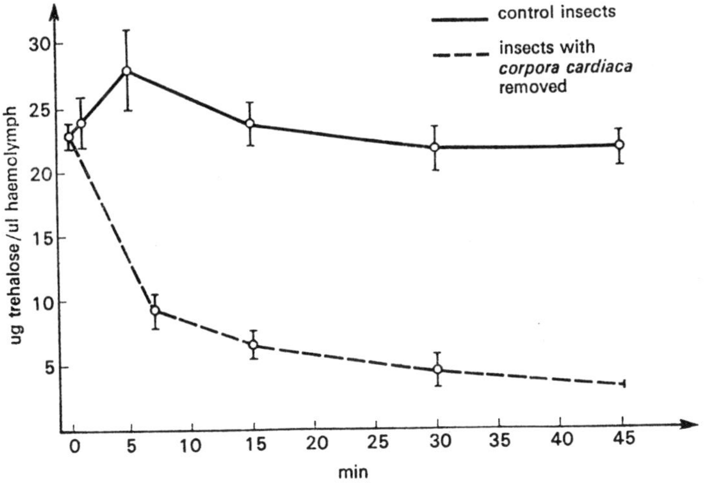

Figure 7.7. Involvement of hyperglycaemic hormone in maintenance of trehalose level in the haemolymph during flight of *Calliphora erythrocephala*. After Vejberg and Normann (1974), with permission.

Gäde (1977) and Ziegler (1979) found that the *corpora cardiaca* of *Manduca sexta* must contain large quantities of the hyperglycaemic hormone, since as little of the extract as 1/100 gland equivalent is sufficient for maximal stimulation of glycogen phosphorylase activity in the fat body (Fig. 7.6). An effective mechanism for raising the trehalose level in the haemolymph is necessary particularly during periods of enhanced activity, such as flying. Experiments with *Calliphora erythrocephala* demonstrated that after removal of the *corpora cardiaca*, the trehalose concentration falls markedly after as little as 15 min of flight (Fig. 7.7); after 45 min the insects were incapable of further flight, although they still had large glycogen reserves in the fat body and haemolymph (Vejberg and Normann, 1974). In control insects, however, the trehalose concentration increased during the first 5 min of flight, and remained at a high level.

7.2.1.2. The hypoglycaemic hormone

The mechanism depressing the glucose level in the haemolymph of insects has been relatively little studied. The existence of a hypoglycaemic hormone was suggested by Dixit and Patel (1964), Ishay *et al.* (1976), Tager *et al.* (1976) and Kramer (1980). The action of this hormone is similar to that of insulin in mammals: extracts of whole fruit flies (*Drosophila melanogaster*) injected into mice produce a decrease in the blood glucose level. Injection of insects with vertebrate insulin has a similar effect (Ishay *et al.*, 1976; Duve, 1978).

Normann (1975) suggested that the hypoglycaemic hormone is produced by the medial neurosecretory cells of the brain in *Calliphora erythrocephala*. This was confirmed by Duve (1978), who demonstrated that removal of cells of the *pars intercerebralis* of *Calliphora erythrocephala* females leads to considerable increases in trehalose and glucose levels in the haemolymph 24 h after treatment. Injection of a homogenate of neurosecretory cells into insects from which the cells had been removed 24 h earlier, however, caused a considerable reduction in the content of these carbohydrates as soon as 4 h after injection. These results suggest that the brains of these insects must contain some substance whose action is similar to that of insulin in vertebrates.

7.2.2. Influence of ecdysone on carbohydrate metabolism

To date there is no evidence for the direct participation of ecdysone in carbohydrate metabolism. Ito and Horie (1957) found that ligation of *Bombyx mori* pupae so as to isolate the prothoracic glands prevents glycogenolysis, whereas implantation of ligated pupae with these glands activates this process. These results were confirmed in experiments with pupae in which metabolic processes were inhibited by removal of the brain. If ecdysone is injected into such pupae, trehalose synthesis is stimulated almost immediately, with simultaneous inhibition of glycogen synthesis (Kobayashi and Kimura, 1967). It seems, however, that the action of ecdysone is indirect. The action of this hormone is principally morphogenetic, and this of course is associated with a high energy requirement, which is satisfied by trehalose. This supposition is confirmed by the results of investigations indicating that ecdysone facilitates trehalose transport to the cells by increasing the permeability of the cell membrane (Jungreis and Wyatt, 1972). The fall in the trehalose level in the haemolymph would stimulate its synthesis by a feedback mechanism, and this is what is observed after the administration of ecdysone.

7.2.3. Octopamine and 5-hydroxytryptamine (5-HT) in carbohydrate metabolism

Robertson and Steele (1973a) detected the presence of octopamine at a high concentration in the abdominal nerve cord of *Periplaneta americana.* In further studies these authors demonstrated that octopamine causes a significant decrease in glycogen content in this tissue (Robertson and Steele, 1973b). Octopamine has a similar influence on the fat body. This effect probably involves the activation of glycogen phosphorylase, since injection of 0.1 mmol/ml of octopamine into the body cavity increases the activity of this enzyme by as much as 48 per cent compared with control results (Robertson and Steele, 1972). Activation of this enzyme most probably involves the participation of cAMP, as demonstrated by Hart and Steele (1973). Similar results were obtained by Nathanson and Greengard (1973) who found that injection of 0.25 mmol/ml of octopamine causes a 7-fold increase in cyclic AMP.

An opposite effect to that of octopamine seems to be produced by 5-HT, the presence of which has been detected in the brain, abdominal cord and *corpora cardiaca* (Gersch *et al.*, 1961; Colhoun, 1963; Muszyńska-Pytel and Cymborowski, 1978). It was noted that this amine exerts an inhibitory influence on the glycogen phosphorylase activity of the nerve cord in *Periplaneta americana* (Hart and Steele, 1973). The mechanism of this inhibition is so far unknown. It is probable that 5-HT acts by depressing the calcium ion concentration in the cell, since it is known that Ca^{++} ions are highly effective in the activation of phosphorylase (Brostrom *et al.*, 1971). Further investigations are necessary.

7.3. Lipid metabolism

The first indication that lipid metabolism in insects is controlled by hormones was the finding of an accumulation of lipids in the fat body of *Melanoplus differentialis* after removal of the *corpora allata* (Pfeiffer, 1945). For the first time attention was drawn to the role of the fat body as a target tissue for the hormone produced by the *CA*. Although many years have elapsed since this discovery, the mechanism by which lipids are transported to the ovaries during growth and maturation of the eggs remains unknown. This is strange, since we already have much information on the transport of proteins (vitellogenins) (see Chapter 5).

7.3.1. The role of JH in lipid metabolism

The results of Pfeiffer's investigations have been confirmed by numerous authors (Steele, 1976). For instance Hill and Izatt (1974) demonstrated that removal of the *CA* from mature *Schistocerca gregaria* females causes an about 3-fold increase in the lipid level in the fat body compared with that in control insects. The same authors also noted that the increased accumulation of lipids after allatectomy does not results from an increased amount of available food, but from changes in the metabolic pathways of the insects. One such change may be in the relative proportions of saturated and unsaturated fatty acids. Stephen and Gilbert (1970) demonstrated that in *Hyalophora cecropia* JH increases the accumulation of saturated fatty acids. These authors advanced an interesting hypothesis explaining the role of JH in this process.

The increased lipid accumulation in the fat body of allatectomized insects suggests either that lipid uptake is partly inhibited, or that lipid synthesis increases. There is much evidence that a deficiency of JH in the haemolymph does not disturb lipid uptake by cells and tissues (Doane, 1961, Walker and Bailey, 1971). It would seem, therefore, that JH acts on the synthesis of lipids rather than on their utilization. Indeed, Morohoshi and Fugo (1972) demonstrated that in allatectomized pupae of the silkworm, the metabolism of glucose into lipids is enhanced. They showed this using glucose labelled with radioactive carbon.

It should be added that JH plays an important role in vitellogenesis, especially in the synthesis of vitellogenins, which are in fact lipoproteins (see Chapter 5). It is therefore difficult to separate the processes of protein and lipid synthesis during vitellogenesis. It is as yet not clear how JH controls two different processes.

7.3.2. Involvement of the adipokinetic hormone in lipid metabolism

In insects capable of flights of long duration (e.g. the migratory locust) lipids are the main source of energy (Weis-Fogh, 1952). The question therefore arises as to the mechanism by which lipid is released from the fat body and transported to the muscles. Chino and Gilbert (1965) demonstrated that lipids are transported from the fat body as complexes of diglycerides with haemolymph proteins. Lipid is released from the fat body under the influence of the adipokinetic hormone (Kammer and Heinrich, 1978). In the locust, trehalose is utilized for the first 5–10 min of flight (Fig. 7.8). When the concentration of this carbohydrate

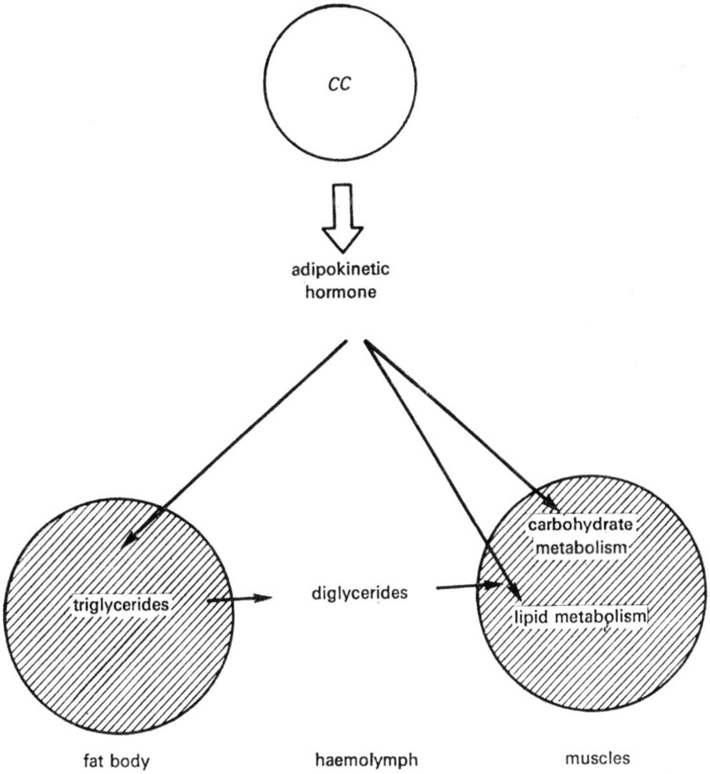

Figure 7.8. Mechanism of action of adipokinetic hormone in the migratory locust. The hormone secreted by *corpora cardiaca* (*CC*) stimulates metabolism of triglycerides in the fat body to diglycerides, which are released into the haemolymph. This hormone also acts on the muscle moving the insect's wings, causing a change from carbohydrate metabolism to lipid metabolism. Further details in text.

in the haemolymph reaches a low level, the *corpora cardiaca* secrete the adipokinetic hormone, leading to the release of diglycerides from the fat body (Beenakkers *et al.*, 1978). This protein hormone has been synthesized in the laboratory (Stone *et al.*, 1976). It probably acts by the intermediary of cyclic AMP. Injection of locusts with 10 pmol of the adipokinetic hormone increases by about threefold the cAMP content of the fat body (Gäde, 1979).

In the working insect muscle the adipokinetic hormone enhances β-oxidation, leading to oxidation of fatty acids (Robinson and Goldsworthy, 1977). These authors postulate direct action of the hormone on the mitochondria of the muscle cells. The mechanism of this action is, however, not known.

It is noteworthy that vertebrate adrenalin when injected into insects has an effect similar to that of the adipokinetic hormone; it releases only free fatty acids from the fat body of *Hyalophora cecropia* however, and not, as in the case of the adipokinetic hormone, diglycerides (Bhaktan and Gilbert, 1968). It is therefore difficult to evaluate the importance of this discovery.

7.3.3. Influence of the diapause hormone of lipid metabolism

The diapause hormone produced by the suboesophageal ganglion of the silkworm (*Bombyx mori*) causes the female to lay diapausing eggs. Hasegawa and Yamashita (1967) noted that such eggs contain much more lipids than nondiapausing eggs. When the source of the diapausing hormone is removed—the suboesophageal ganglion of pupae that contain nondiapausing eggs—the lipid content in the ovaries diminishes, while in the fat body it increases (Morohoshi and Fugo, 1971). Opposite results are obtained after implantation of the oesophageal ganglion into pupae. It seems, therefore, that the diapause hormone facilitates fatty acid uptake by the ovaries, although it may also stimulate synthesis of these acids. This suggests an increased metabolism of ^{14}C-glucose to ovarian lipids in the presence of the diapause hormone (Morohoshi and Fugo, 1971). At the same time, however, increased activity of trehalase is observed in the ovaries (Yamashita *et al.*, 1972). It would seem, therefore, that the influence of the diapause hormone on lipid synthesis is not a direct one.

7.4. Amino acid metabolism

Best known to date is the hormonal regulation of the metabolism of tyrosine, which after transformation to quinone participates in cuticle tanning and sclerotization. Pryor (1940) was the first to call attention to this process. Later Brunet (1969) presented the series of chemical reactions involved.

7.4.1. Role of ecdysone in tyrosine metabolism

Many investigators have drawn attention to the fact that before each moult the tyrosine concentration in the haemolymph increases markedly (Fraenkel and Rudall, 1947; Patterson, 1957; Post and De Jong, 1973). The appearance of tyrosine at these times in insect development may be brought about by ecdysone, since moulting is preceded by an increase in the concentration of this hormone in the haemolymph (see Chapter 4). In *Sarcophaga bullata* a particularly large increase in the tyrosine level was noted before the imaginal moult (Seligman *et al.*, 1969). What is more, injection of young larvae with ecdysone increases the tyrosine concentration in the haemolymph and causes premature tanning of the cuticle (Bodnaryk, 1971). This process involves the following transformation: tyrosine → DOPA—dopamine → N-acetyl-dopamine. The last step, the transformation of N-acetyl-dopamine into pigment, involves phenol oxidase, which requires an activating factor. It seems that such a factor is produced in response to the action of ecdysone, since removal of the ring gland (containing the prothoracic glands) in *Calliphora erythrocephala* produces a decrease in phenol oxidase activity. On the other hand, injection of ecdysone into *Calliphora erythrocephala* larvae after removal of this gland maintains synthesis of the activating agent, and no decrease in phenol oxidase activity is observed (Karlson and Schweiger, 1961).

7.4.2. Involvement of JH in uric acid production

Thomas and Nation (1966) demonstrated that removal of the *corpora allata* in *Periplaneta americana* brings about a 60 per cent fall in the rate of protein synthesis rate in various tissues. This reduction in protein synthesis leads to a great accumulation in the haemolymph of amino acids, from which uric acid is then synthesized (Coles, 1965; Mordue and Goldsworthy, 1973).

It would seem, therefore, that JH participates only indirectly in the regulation of uric acid synthesis in insects.

7.5. Summary

Insects are an excellent model for research on the hormonal regulation of metabolic processes, especially now that many insect hormones have been isolated and investigated from the physico-chemical aspect. The availability of numerous analogues of these hormones in large quantities greatly facilitates such studies. This is particularly important in the case of the hormonal regulation of respiratory metabolism in insects. The detection of the phenomenon of hypermetabolism in *Dermestes vulpinus* larvae allows a deeper insight into the regulation of respiratory processes at the molecular level, since in this case oxygen uptake is not directly associated with the morphogenetic changes that normally accompany the action of juvenile hormone in a number of groups of insects.

The results of numerous investigations indicate that JH is indirectly involved in the regulation of the rate of oxygen uptake. Allatectomy nearly always leads to a considerable depression in respiratory metabolism. On the other hand, the hormone(s) produced by the *corpora cardiaca* stimulate this process.

The detection of the adipokinetic hormone produced by the *CC* and the understanding of its chemical structure is of paramount importance in studies on the lipid metabolism of insects. Now that synthetic adipokinetic hormone is available, we may expect in the near future many new discoveries concerning the mechanism of its action.

7.6. References

Beenakkers A. M. Th., Van der Horst O. J., Van Marrewijk W. J. A. 1978. Regulation of release and metabolic function of the adipokinetic hormone in insects. [In:] *Comparative Endocrinology* P. J. Gaillard and H. H. Boer (ed.), pp. 445–448. Elsevier, North-Holland Biomedical Press, Amsterdam, Oxford, New York.

Bhaktan N. M. G., Gilbert L. I. 1968. Effects of some vertebrate hormones on lipid mobilization in the insect fat body. *Gen. Comp. Endocrinol.*, **11**; 186–197.

Bodnaryk R. P. 1971. Effect of exogenous moulting hormone (Ecdysterone) on β-alanyl-L-tyrosine metabolism in the larva of the fly *Sarcophaga bullata* Parker. *Gen. Comp. Endocrinol.*, **16**; 363–368.

Brostrom C. O., Hunkeler F. L., Krebs E. G. 1971. The regulation of skeletal muscle phosphorylase kinase by Ca^{++}. *J. Biol. Chem.*, **246**; 1961–1967.

Brunet P. C. J. 1963. Tyrosine metabolism in insects. *Amer. N.Y. Acad. Sci.*, **100**; 1020–1034.

Chino H., Gilbert L. J. 1965. Lipid release and transport in insects. *Biochim. Biophys. Acta*, **98**; 94–110.

Clark K. U., Baldwin R. W. 1960. The effect of insect hormones and 2,4-dinitrophenol on the mitochondrion of *Locusta migratoria* L. *J. Insect Physiol.*, **5**; 36–46.

Coles G. C. 1965. Studies of the hormonal control of metabolism in *Rhodnius prolixus* Stal. I. The adult female insect. *J. Insect Physiol.*, **11**; 1325–1330.

Colhoun E. M. 1963. Synthesis of 5-hydroxytryptamine in the American cockroach. *Experientia*, **19**; 9–10.

Conradi-Larsen E. M. 1970. Influence of neurosecretion and nutrition on O_2 consumption in the bug *Oncopeltus*. *J. Insect Physiol.*, **16**; 471–482.

Dixit P. K., Patel N. G. 1964. Insulin-like activity of larval foods of the honeybee. *Nature*, **207**; 4928.

Doane W. W. 1961. Developmental physiology of the mutant female sterile (2) adipose of *Drosophila melanogaster*. III. *Corpus allatum* complex and ovarian transplantations. *J. exp. Zool.*, **146**; 275–298.

Duve H. 1978. The presence of a hypoglucemic and hypotrehalocemic hormone in the neurosecretory system of the blowfly *Calliphora erythrocephala*. *Gen. Comp. Endocrinol.*, **36**; 102–110.

Firstenberg D. E., Silhaček D. L. 1973. Juvenile hormone regulation of oxidative metabolism in isolated insect mitochondria. *Experientia*, **29**; 1420–1422.

Fraenkel G., Rudall K. M. 1947. The structure of insect cuticles. *Proc. R. Soc. B.*, **134**; 113–143.

Gäde G. 1977. Effect of *corpus cardiacum* extract on cyclic AMP concentration in the fat body of *Periplaneta americana*. *Zool. Jb. Physiol.*, **81**; 245–249.

Gäde G. 1979. Studies on the influence of synthetic adipokinetic hormone and some analogues on cyclic AMP levels in different arthropod systems. *Gen. Comp. Endocrinol.*, **37**; 122–130.

Gersch M., Fischer F., Unger H., Kabitzer W. 1961. Presence of 5-HT in the nervous system of *Periplaneta americana*. *Z. Naturforsch.*, **161**; 351–352.

Gilbert L. J. 1967. Lipid metabolism and function in insects. *Adv. Insect Physiol.*, **4**; 70–208.

Goldsworthy G. J. 1969. Hyperglicaemic factors from the *corpus cardiacum of Locusta migratoria*. *J. Insect Physiol.*, **15**; 2131–2140.

Hart D. E., Steele J. E. 1973. The glycogenolytic effect of the *corpus cardiacum* on the cockroach nerve cord. *J. Insect Physiol.*, **19**; 927–939.

Hasegawa K., Yamashita O. 1967. Control of metabolism in the silkworm pupal ovary by the diapause hormone. *J. Sericult. Sci. Japan*, **36**; 297–301.

Hill L., Izatt M. E. 1974. The relationship between *corpora allata* and fat body and hemolymph lipids in the adult female desert locust. *J. Insect Physiol.*, **20**; 2143–2156.

Ishay J., Gritten S., Galun R., Dorn M., Leron Z. 1976. The presence of insulin and some effects of exogenous insulin on *Hymenoptera* tissues and body fluids. *Comp. Biochem. Physiol.*, **54A**; 203–206.

Ito T., Horie Y. 1957. Glycogen in the ligated silkworm pupa (*Bombyx mori*). *Nature*, **179**; 1136.

Jungreis A. M., Wyatt G. R. 1972. Sugar release and penetration into insect fat body: relations to regulation of haemolymph trehalose in developing stages of *Hyalophora cecropia*. *Biol. Bull.*, **143**; 367–391.

Kammer A. M., Heinrich B. 1978. Insect flight metabolism. *Adv. Insect Physiol.*, **13**; 133–228.

Karlson P., Schweiger A. 1961. Zum Tyrosinstoffwechsel der Insekten. IV. Das Phenoloxydase-System von *Calliphora* und seine Beeinflussung durch das Hormon Ecdyson. *Z. Phys. Chem.*, **323**; 199–210.

Keeley L. 1970. Insect fat body mitochondria: endocrine and age effects on respiratory and electron transport activities. *Life Sci.*, **9**; 1003–1011.

Keeley L. 1971. Endocrine effects on the biochemical properties of fat body mitochondria from the cockroach, *Blaberus discoidalis*. *J. Insect Physiol.*, **17**; 1501–1515.

Kobayashi M., Kimura S. 1967. Action of ecdysone on the conversion of ^{14}C-glucose in dauer pupa of the silkworm, *Bombyx mori*. *J. Insect Physiol.*, **13**; 545–552.

Kramer K. J. 1980. Insulin-like and glucagon-like hormones in insects. [In:] Neurohormonal

Techniques in Insects T. A. Miller (ed.), pp. 116–136, Springer-Verlag, New York, Heidelberg, Berlin.

Kryspin J. 1977. *Wpływ pochodnych hormonu juwenilnego na metabolizm oddechowy owadów* [*Influence of juvenile hormone derivatives on insect respiratory metabolism*]. Ph. D. thesis, Univ. of Warsaw.

Kryspin-Sørensen J., Gelbič J., Sláma K. 1977. Juvenoid action on the total body metabolism in larvae of noctuid moth. *J. Insect Physiol.*, **23**; 531–535.

Lafont-Cazal M., Roussel J. P. 1971. Modifications metaboliques après ablation simultanée des *corpora allata* et des *corpora cardiaca* chez *Schistocerca gregaria* et *Locusta migratoria. C. R. Acad. Sci. Paris*, **273**; 2613–2615.

L'Hélais C. 1954. Étude du métabolisme basal chez le phasme *Dixippus morosus* après ablation des *corpora allata. C. R. Acad. Sci. Paris*, **239**; 778–780.

L'Hélais C. 1964. Aspects chimiques des méchanismes chez les insectes. *Ann. Biol.*, **3**; 1–31.

Lüscher M., Leuthold R. 1965. Über die hormonale Beeinflüssung des respiratorischen Stoffwechsels bei der Schabe *Leucophaea maderae* (F). *Rev. Suisse Zool.*, **72**; 618–623.

Minks A. K. 1967. Biochemical aspects of juvenile hormone action in the adult *Locusta migratoria. Arch. Neer. Zool.*, **17**; 175–249.

Mordue W., Goldsworthy G. J. 1973. Transaminase levels and uric acid production in adult insects. *Insect Biochem.*, **3**; 419–427.

Morohoshi S., Fugo H. 1971. The control of growth and development in *Bombyx mori.* X. Effect of the suboesophageal ganglion hormone on the incorporation of ^{14}C glucose into the lipids of the fat body, haemolymph and ovary. *Proc. Imp. Acad. Japan*, **47**; 416–421.

Morohoshi S., Fugo H. 1972. The control of growth and development in *Bombyx mori.* XI. Effect of the *corpus allatum*, suboesophageal ganglion, brain and ecdysterone on the incorporation of ^{14}C-U-glucose into pupal fat body and ovary lipids. *Proc. Imp. Acad. Japan*, **48**; 127–132.

Müller H. P., Engelmann F. 1968. Studies on the endocrine control of metabolism in *Leucophaea maderae* (*Blattaria*). II. Effect of the *corpora cardiaca* on fat body respiration. *Gen. Comp. Endocrinol.*, **11**; 43–50.

Muszyńska-Pytel M., Cymborowski B. 1978. The role of serotonin in regulation of the circadian rhythms of locomotor activity in the cricket (*Acheta domesticus* L.). II. Distribution of serotonin and variations in different brain structures. *Comp. Biochem. Physiol.*, **59C**; 17–20.

Nathanson J. A., Greengard P. 1973. Octopamine-sensitive adenylate cyclase: evidence for a biological role of octopamine in nervous tissue. *Science*, **180**; 303–310.

Niemierko W. 1956. Badania nad metabolizmem *Galleria mellonella* i *Bombyx mori* L. [Investigations on the metabolism of *Galleria mellonella* and *Bombyx mori* L.]. *Acta Biochem. Pol.*, **3**; 627–647.

Normann T. C., Duve H. 1969. Experimentally induced release of a neurohormone influencing hemolymph trehalose level in *Calliphora erythrocephala* (*Diptera*). *Gen. Comp. Endocrinol.*, **12**; 449–459.

Normann T. C. 1975. Neurosecretory cells in insect brain and production of hypoglycaemic hormone. *Nature*, **254**; 259–261.

Patterson D. S. P. 1957. Qualitative and quantitative changes observed in the free amino nitrogen fraction of *Tenebrio molitor* pupal tissues during metamorphosis. *Biochem. J.*, **65**; 729–735.

Pfeiffer J. W. 1945. Effect of the *corpora allata* on the metabolism of adult females grasshoppers. *J. exp. Zool.*, **99**; 183–233.

Post L. C., DeJong B. J. 1973. Bursicon and metabolism of tyrosine in the moulting cycle of *Pieris larvae. J. Insect Physiol.*, **19**; 1541–1546.

Pryor M. G. M. 1940. On the hardening of the cuticle of insects. *Proc. R. Soc. B.*, **128**; 393–407.

Robertson H. A., Steele J. E. 1972. Activation of insect nerve cord phosphorylase by octopamine and adenosine 3′,5′monophosphate. *J. Neurochem.*, **19**; 1603–1606.

Robertson H. A., Steele J. E. 1973a. Octopamine in the insect central nervous system: distribution, biosynthesis and possible physiological role. *J. Physiol.*, **237**; 34–35.

Robertson H. A., Steele J. E. 1973b. Effect of monophenolic amines on glycogen metabolism in the nerve cord of the American cockroach *Periplaneta americana. Insect Biochem.*, **3**; 53–59.

Robinson N. L., Goldsworthy G. 1977. A possible site of action for adipokinetic hormone on the flight muscle of locusts. *J. Insect Physiol.*, **23**; 153–158.

Sehnal F., Sláma K. 1966. The effect of *corpus allatum* hormone on respiratory metabolism during larval development and metamorphosis of *Galleria mellonella* L. *J. Insect Physiol.*, **12**; 1333–1342.

Sehnal F., Metwally M. M., Gelbič J. 1976. Reaction and sensitivity of Noctuid moths to juvenoids. *Z. angew. Entomol.*, **78**; 115–121.

Seligman M. Friedman S., Fraenkel G. 1969. Bursicon mediation of tyrosine hydroxylation during tanning of the adult cuticle of the fly *Sarcophaga bullata. J. Insect Physiol.*, **15**; 553–562.

Silhaček D. L., Firstenberg D. E., Kohl K. 1974. Juvenile hormone action on the intermediary metabolism in indian meal moth larvae. *Proc. First Intern. Working Confer. Stored-Entomol., Savannah, Georgia USA*, 669–674.

Sláma K. 1964. Hormonal control of respiratory metabolism during growth, reproduction and diapause in female adults of *Pyrrhocoris apterus. J. Insect Physiol.*, **10**; 283–304.

Sláma K. 1965. Effect of hormones on growth and respiratory metabolism in the larvae of *Pyrrhocoris apterus* L. (*Hemiptera*). *J. Insect Physiol.*, **11**; 113–122.

Sláma K., Romaňuk M., Šorm F. 1974. *Insect hormones and bioanalogues*. Springer-Verlag, Wien, New York.

Sláma K., Hodková M. 1975. Insect hormones and bioanalogues: their effect on respiratory metabolism in *Dermestes vulpinus* F. (*Coleoptera*). *Biol. Bull.*, **148**; 320–332.

Sláma K. 1976. Some old concepts and new findings in hormonal control of insect metamorphosis. *J. Insect Physiol.*, **21**; 921–955.

Sláma K., Kryspin-Sørensen J. 1979. Hypermetabolic response induced by juvenile hormone analogues in insects. *Z. Naturforsch.*, **34c**; 599–607.

Sroka P., Barth R. H. 1976. Hormonal control of diglyceride metabolism during vitellogensis in the cockroach *Eublaberus posticus. J. Insect Physiol.*, **22**; 951–954.

Steele J. E. 1961. Occurrence of a hyperglycaemic factor in the *corpus cardiacum* of an insect. *Nature*, **192**; 680–681.

Steele J. E. 1963. The site of action in insect hyperglycaemic hormone. *Gen. Comp. Endocrinol.*, **3**; 46–52.

Steele J. E. 1976. Hormonal control of metabolism in insects. *Adv. Insect Physiol.*, **12**; 239–323.

Stegawee D., van Kammen-Werthen A. R. 1962. Respiratory chain metabolism in the Colorado potato beetle. I. Respiration and oxidative phosphorylation in sarcosomes from active beetle. *J. Insect Physiol.*, **8**; 117–126.

Stephen W. E., Gilbert L. J. 1970. Alterations in fatty acids composition during the metamorphosis of *Hyalophora cecropia*: correlations with juvenile hormone titre. *J. Insect Physiol.*, **16**; 851–864.

Stone J. V., Mordue W., Batley K. E., Morris H. R. 1976. Structure of locust adipokinetic hormone: a neurohormone that regulates lipid utilization during flight. *Nature*, **263**; 207–210.

Tager H. S., Markese J., Speirs R. D., Kramer K. J. 1975. Glucagon-like immunoreactivity in insect *corpus cardiacum*. *Nature*, **254**; 707–708.

Tager H. S., Markese J., Kramer K. J., Speirs R. D., Childs C. N. 1976. Glucagon-like and insulin-like hormones of the insect neurosecretory system. *Biochem. J.*, **156**; 515–520.

Thomas K. K., Nation J. L. 1966. RNA protein and uric acid content of body tissues of *Periplaneta americana* (L.) as influenced by *corpora allata* during ovarian development. *Biol. Bull.*, **130**; 442–449.

Vejberg K., Normann T. C. 1974. Secretion of hyperglycaemic hormone from the *corpus cardiacum* of flying blowflies, *Calliphora erythrocephala*. *J. Insect Physiol.*, **20**; 1189–1192.

Walker P. R., Bailey E. 1971. Effect of allatectomy on fat body lipid metabolism of the male desert locust during adult development. *J. Insect Physiol.*, **17**; 813–821.

Weis-Fogh T. 1952. Fat combustion and metabolic rate of flying locusts (*Schistocerca gregaria* Forskal). *Phil. Trans. R. Soc. Ser. B.*, **237**; 1–36.

Wiens A. W., Gilbert L. J. 1967. Regulation of carbohydrate mobilization and utilization in *Leucophaea maderae*. *J. Insect Physiol.*, **13**; 779–794.

Wojtczak L. 1952. Studies on the biochemistry of the wax moth—10. Respiratory enzymes of the larva. *Acta Biol. exp. Wars.*, **16**; 199–222.

Wojtczak L., Wojtczak A. B. 1960. Uncoupling of oxidative phosphorylation and inhibition of ATP-Pi exchange by a substance from insect mitochondria. *Biochem. Biophys. Acta*, **39**; 277–286.

Wojtczak A. B. 1969. Utlenianie bursztynianu w mitochondriach [Succinate oxidation in the mitochondria]. *Post. Bioch.*, **15**; 339–356.

Wojtczak L., Bogucka K., Sarzała M. G., Załuska H. 1969. Effect of fatty acids on energy metabolism and the transport of adenine nucleotides in mitochondria and other cellular structures. *FEBS Symposium*, **17**; 79–93.

Wyatt G. R. 1967. The biochemistry of sugars and polysaccharides in insects. *Adv. Insect Physiol.*, **4**; 287–360.

Yamashita O., Hasegawa K., Seki M. 1972. Effect of the diapause hormone on trehalase activity in pupal ovaries of the silkworm, *Bombyx mori* L. *Gen. Comp. Endocrinol.*, **18**; 515–523.

Ziegler R. 1979. Hyperglycaemic factors from the *corpora cardiaca* of *Manduca sexta* (L.) (*Lepidoptera*: *Sphingidae*). *Gen. Comp. Endocrinol.*, **39**; 350–357.

Chapter 8

Hormones and insect behaviour

8.1. General characteristics of hormonal regulation of insect behaviour

There is now an extensive bibliography on the involvement of hormones in the regulation of various aspects of insect behaviour (Markl, 1974; Truman and Riddiford, 1974, 1977; Truman 1978a). Experimental evidence supporting the involvement of hormones can be classified into four main groups:

1. the finding of a correlation between the appearance in the haemolymph of a given hormone and a change in the behaviour of the insect;
2. premature appearance of a certain behaviour after application of a hormone or implantation of active endocrine gland;
3. cessation of a typical behaviour after removal of a hormone source;
4. the possibility of reproducing a behavioural pattern that occurred before removal of a hormone source by reimplantation of the endocrine gland secreting the hormone.

The first type of evidence seems the weakest and would require additional experimental support. The correlation found may be accidental, and it is hardly possible to distinguish on this basis between the cause and the consequence. It may be true that the hormone produces a change in behaviour, but the opposite is also possible: a change in behaviour due to external stimuli may lead to hormonal changes.

The third type of evidence also leaves room for doubts; removal of the secretory cells of the brain, for instance, may be accompanied by the destruction of other neurons, and this may evoke fundamental changes in the insect's behaviour.

None of the types of evidence listed above, considered separately, warrants conclusions about the hormonal basis of the behaviour studied, and they should as far as possible be considered jointly.

Let us now consider, from a theoretical point of view, the possible effects of hormones on insect behaviour. Since insects undergo metamorphosis during development, hormones can cause atrophy of some motoneurons and stimulate the growth of others (Truman, 1973). It was noted in *Manduca sexta*, for instance, that during the imaginal moult many motoneurons and interneurons disappear (Fig. 8.1). The same eclosion hormone thus stimulates two different types of behaviour: the premoulting behaviour of the pupa (characteristic movement of the abdomen) and imaginal behaviour (e.g. stretching of the wings). Investigations of this type are extremely important because they reveal as it were the anatomical substrate for the action of hormones. They do not, however, explain how a particular type of insect behaviour is switched on or off. It is also difficult to explain the fact that the change of behaviour may occur, and usually does occur, without noticeable morphological changes in the nervous system. The second possible mode of action of hormones involves a change in the sensitivity of the nervous system and thus, of the whole organism to external stimuli. From

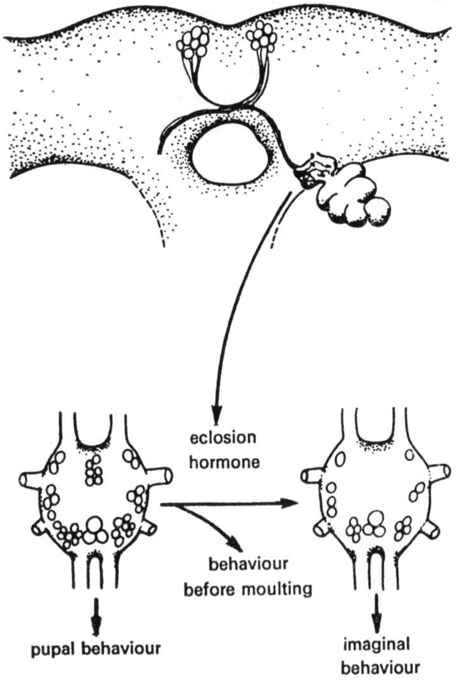

Figure 8.1. Eclosion hormone is produced by the medial neurosecretory cells of the brain and released into the haemolymph via the *corpora cardiaca*. It acts on the abdominal ganglia, stimulating premoulting behaviour. During the imaginal moult many neurons responsible for behaviour during the pupal stage disappear completely. After Truman (1973), with permission.

the technical point of view it is extremely difficult to obtain satisfactory evidence for such a mechanism of hormone action, and there have been few experimental studies in this field. However, one example is the change in the stereotypic walking movements of the cockroach after injection of *corpora cardiaca* homogenate (Ozbas and Hodgson, 1958). These authors demonstrated that extracts of the *CC* produce a change in the spontaneous bioelectric activity of the interneurons in the nerve cord *in vitro*.

Another way in which hormones affect insect behaviour is through changes in various peripheral organs, particularly the proprioreceptors. Such changes may occur through direct action of hormones on the central nervous system of the insect (Ždárek and Sláma, 1968). We will devote much space to these problems when discussing the different patterns of insect behaviour during development and metamorphosis.

In general, insect hormones regulating behaviour may be divided into two main groups: those that trigger a particular behaviour, and those that modify behaviour (Truman and Riddiford, 1974).

The triggering hormones are characterized by the fact that they are capable of "mobilizing" the behaviour concerned very rapidly (within several minutes) without the need for stimuli such as those from the environment (Fig. 8.2 A). In this respect the action of hormones resembles that of pheromones (Wilson and Bossert, 1963), since both hormones and pheromones trigger a particular stereotypic behaviour. An example of this kind of hormone action is the influence of *corpora cardiaca* hormones on the behaviour of silkworm females when calling males (Riddiford and Williams' 1971). In the same species, eclosion hormone acts on the central nervous system (CNS), triggering a stereotypic programme of abdominal movements in the pupa, usually within 10–15 min (Truman and Sokolove, 1972; Truman, 1976).

The action of modifying hormones is much slower, and consists mainly in producing changes in the sensitivity of the nervous system to stimuli. Before this type of hormone acts, a given stimulus causes one type of behavioural reaction, whereas after release of the hormone the same stimulus produces a completely different response (Fig. 8.2 B).

A typical example of a change in behaviour of the insect under the influence of a hormone is the effect of JH on the susceptibility of *Gomphocerus rufus* females to the calling of the male (Loher and Hüber, 1966). Immediately after the imaginal moult the female refuses to let the male approach her, and is aggressive towards him, whereas after JH administration the calling of the male is favourably accepted and copulation takes place (Loher, 1966). The eclosion hormone acts in a similar

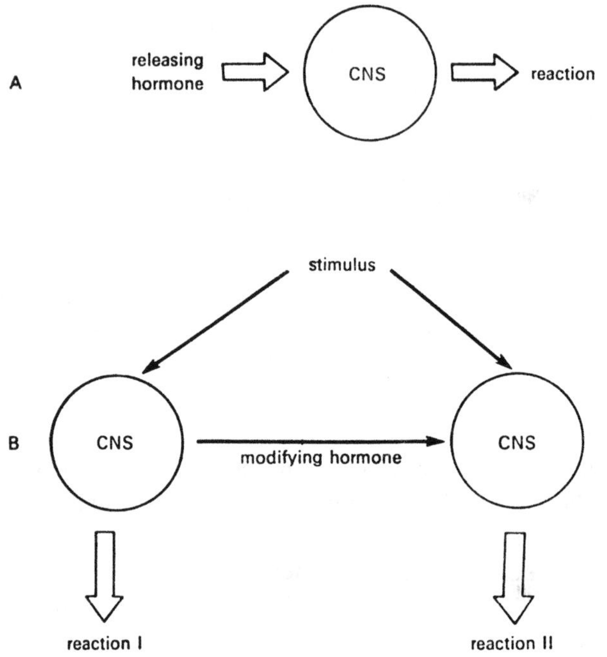

Figure 8.2. Action of eclosion hormone and modifying hormone on the insect central nervous system (CNS). Further details in text. After Truman and Riddiford (1974), with permission.

way on the reproductive behaviour of *Antherea pernyi.* In this species copulation occurs 15 min after the imaginal moult if the insects are placed in a cage with dim light (Riddiford, 1970). For copulation to take place, however, the eclosion hormone must act on the male, making him sensitive to the pheromone produced by the female. If the imago is artificially removed from the pupa too soon before natural eclosion, preventing secretion of the eclosion hormone, the male does not react at all to the pheromone, even under optimal environmental conditions. This lack of sensitivity to pheromones is regulated by the nervous system, and does not result from blocking of the receptors present on the insect's antennae, since electrophysiological studies (electroantennograms) confirm their reaction to pheromones. Thus, we see that lack of the modifying hormone renders the insect's nervous system incapable of initiating the appropriate responses to stimuli from the environment, so that, in this case, reproductive behaviour is not evoked.

It may be said, therefore, that hormones determine the ability of the insect to perform particular types of behaviour. This occurs mainly by the preparation of an appropriate structural substrate (Hoyle, 1970) or by producing an appropriate "mood" in the insect (Loher and Hüber, 1966; Žďárek, 1970, Barth and Lester, 1973; Žďárek and Kontew, 1975; Žďárek, 1981). Hormones also determine when a particular kind of behaviour can be initiated so that it is fully synchronized with the appropriate environmental conditions (Truman, 1971).

8.2. Hormonal regulation of reproductive behaviour in insects

The behaviour of insects associated with their reproduction has an abundant literature, and it is possible to discuss here only a few of its aspects. The pheromones that play an essential role in the regulation of reproductive behaviour (Wilson, 1971; Birch, 1974) will not be considered. As ectohormones, these substances are outside the scope of endocrinology. As we shall see later, the reproductive behaviour of insects is also influenced by the hormones that regulate growth and metamorphosis.

Female insects, especially *Hymenoptera* and *Lepidoptera*, copulate almost immediately after eclosion from the pupa. In extreme cases copulation occurs even before emergence from the pupa, as in *Opifex fuscus* (*Diptera*) and in some species of *Clunioninae*. Females of the cockroach *Diploptera punctata* copulate whilst their wings are still soft and incompletely extended. The mechanism of hormonal regulation of reproductive behaviour in these insects is not known since the imago emerges from the pupa as a completely mature insect capable of reproduction.

A kind of maturation process occurs in insect species in which copulation occurs a relatively long time after the imaginal moult. Before copulation cricket males exhibit a characteristic behaviour associated with the presence of a spermatophore in their reproductive organs (Dumortier, 1963). There is much data indicating, on the one hand, that such behaviour is dependent on nervous information about the presence of the spermatophore, since interrupting the transmission of this information to the brain by transsecting the nerve cord abolishes the behaviour (Hüber, 1955). On the other hand, there is much evidence that this behaviour is not dependent on the gonads, since castration does not abolish it (Engelmann, 1970).

In some female insects there is a rather long period after the imaginal moult during which the male is not allowed to approach for copulation. Thus, a kind of sexual maturation takes place. What is the nature of this process ? It seems to depend directly on the physiological state of the female gonads. In some species, however, removal of the ovaries does not abolish this characteristic female reproductive behaviour, as observed in *Leucophaea maderae* (Engelmann, 1960) or in *Pyrrhocoris apterus* (Ždárek, 1971). It was demonstrated in *Leucophaea maderae* that juvenile hormone participates in the regulation of the receptivity of the female with respect to the male, and that this action is mediated by the central nervous system (Engelmann and Barth, 1968). In *Galleria mellonella*, however, JH is not involved in the regulation of reproductive behaviour (Röller *et al.*, 1963). It would seem, therefore, that JH may be only one of many possible factors controlling this process. Another factor influencing the reproductive behaviour of female insects is the presence of spermatozoa in the spermatheca. Loher (1966) demonstrated that removal of the spermatophore of the cricket (*Gomphocerus rufus*) female immediately after copulation (before the spermatozoa are released) leads to repeated copulation. Transection of the nerves serving the spermatheca or transection of the nerve cord causes copulation to be repeated several times, notwithstanding the presence of the spermatophore in the spermatheca. As demonstrated in the same species, a necessary condition for normal reproductive behaviour is activity of the *corpora allata*, i.e. the presence of a particular JH level in the haemolymph of females (Loher and Hüber, 1966).

It has been demonstrated in many species that a certain minimal concentration of juvenile hormone is required to enable the female to produce the sex hormone that attracts males (Barth, 1968). Removal of the *corpora allata* from female cockroaches (*Byrsotria fumigata*) causes a loss of the ability to secrete sex pheromones; reimplantation of the *CA* restores secretion.

On the basis of this review of the literature, it seems that juvenile hormone is the main hormone influencing the reproductive behaviour of insects. Its mechanism of action is not yet quite clear, but one may suppose that JH can act on reproductive behaviour either directly or indirectly by influencing the production and release of sex pheromones. Further intensive research in this sphere is necessary.

8.3. Hormonal regulation of insect behaviour during development and metamorphosis

8.3.1. Control of larval behaviour

As well as regulating the reproductive behaviour of adults, hormones play an important role in the control of earlier developmental stages. This particularly concerns certain larval instars and the pupa.

8.3.1.1. Hormonal control of feeding behaviour in larvae

Larval stages, especially the final instar, are characterized by periods of very intensive feeding. This is invariably associated with a high level of juvenile hormone in the haemolymph (see Chapter 4). It seems, therefore, that this hormone may stimulate feeding. This is confirmed by the results of investigations in various insects in which the JH level in the haemolymph was artificially increased (see Sláma *et al.*, 1974). In nearly all cases, administration of JH or one of its analogues to final-instar larvae caused considerable lengthening of the feeding period, or even the appearance of additional larval stages. The duration of the feeding period and the accompanying increase in body weight in *Laspeyresia pomonella* are controlled by JH. Larvae of this species reared under short-day conditions have a relatively high level of the hormone, and feed for a much longer period than those reared under long days (Sieber and Benz, 1978). These authors also found that administration of a JH analogue greatly prolongs the feeding period

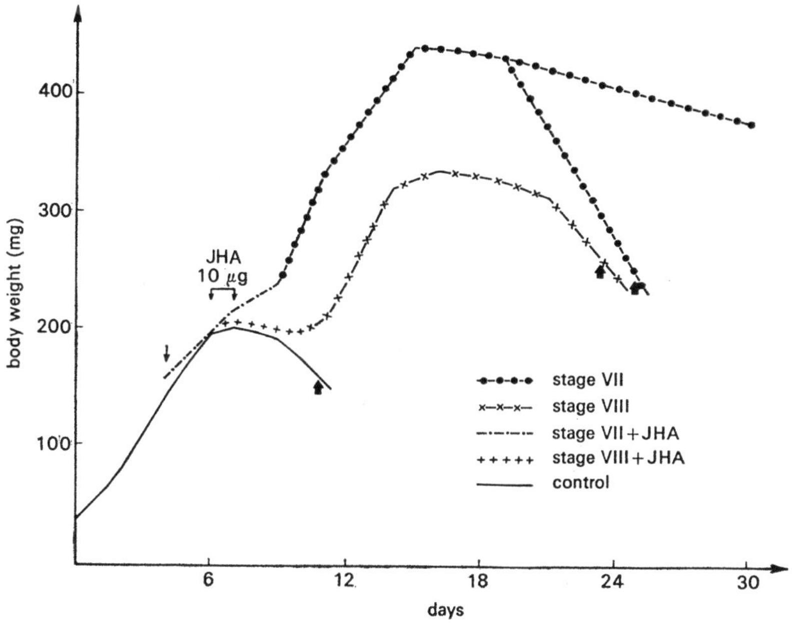

Figure 8.3. Changes in body weight of final-instar larvae of *Galleria mellonella* after application of 10 μg of juvenile hormone analogue (ZR-512). Note considerable increase in feeding time and the additional larval stage (VIII) in some insects. Thin arrow indicates time of moult to last larval instar. Thick arrows indicate times of pupation of larvae. JHA—juvenile hormone analogue. After Ciemior *et al.* (1979), with permission.

in both short-day and long-day larvae. This prolongation is possible only when the hormone is applied during the period when the larvae are feeding. On the other hand, JH does not exert any influence on this process when feeding ceases. This is proved by the fact that at the moment of cessation of feeding a new type of behaviour is switched on, and the preceding behaviour is simultaneously blocked.

Administration of the JH analogue ZR-512 to final instar larvae of *Galleria mellonella* leads to additional larval moults or greatly prolongs the feeding period (Ciemior *et al.*, 1979). Prolonged development of larvae is accompanied by a tremendous increase in body weight (Fig. 8.3). A decrease in the level of JH in the haemolymph of the larvae causes the cessation of feeding and the start of the larval moult or metamorphosis. In the moulting period a considerable inhibition of locomotor activity of the larvae is usually observed. It would seem that this is caused by the appearance of ecdysone in the haemolymph. Such a suggestion was advanced in the case of larval moulting in the locust (Haskell and Moorhouse, 1963). These authors presented results indicating an inhibitory action of ecdysone on the central nervous system of the larva. Further investigations in this respect are, however, necessary.

8.3.1.2. Hormonal control of larval behaviour before metamorphosis

Towards the end of the last larval instar a drastic change of behaviour occurs. The larvae cease to feed and begin an intensive search for a suitable place to spin the cocoon or to build a special pupation chamber in which metamorphosis will occur. In many insects the beginning of the period of wandering is connected with corresponding morphological changes. In *Manduca sexta*, for instance, the pulsating heart becomes visible when the dorsal vessel is exposed (Safranek *et al.*, 1980). The same is noted in *Spodoptera littoralis* (Cymborowski and Stolarz, 1979). The appearance of these changes requires the presence of ecdysone in the haemolymph, but the absence of juvenile hormone. It is not certain whether the initiation of changes in behaviour before metamorphosis results from the fall in JH level in the heamolymph or from the appearance of ecdysone. It seems likely that the decrease in JH level initiates these changes.

As demonstrated by Bounhiol (1938) and Piepho (1950), removal of the *corpora allata* from young *Bombyx mori* larvae leads to premature switching on of a programme of behaviour associated with the spinning of the cocoon, and such larvae produce tiny cocoons. Implantation of *CA* or administration of JH (Riddiford, 1972) greatly retards the switching on of this programme. Similar results were obtained in *Pieris brassicae* larvae (Benz, 1973).

Although a fall in the JH level seems to be necessary for the change from behaviour associated with feeding to the behaviour that precedes metamorphosis, it seems, however, to be insufficient. Lounibos (1974) demonstrated that such a change in behaviour can be evoked prematurely in young larvae of *Hyalophora cecropia* by infusion of ecdysone into the haemolymph. The same may be achieved

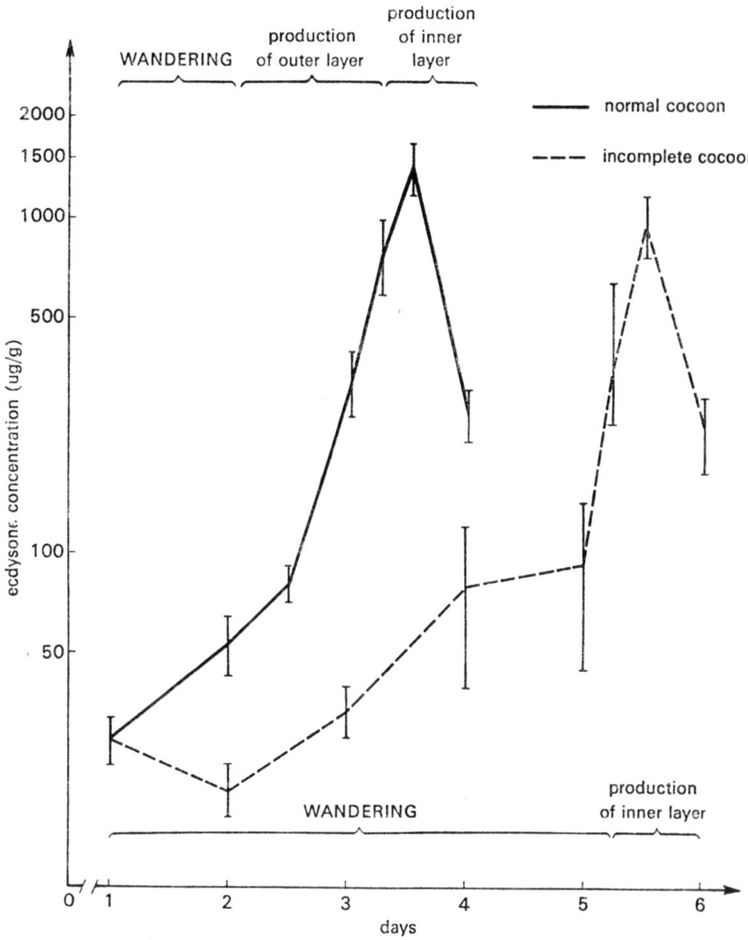

Figure 8.4. Changes in 20-dihydroecdysterone concentration in the bodies of final-instar *Ephestia kuehniella* larvae spinning normal cocoons and those not spinning the outer cocoon layer. Note the persistence of a low 20-dihydroecdysterone level in larvae not spinning the second layer; however, the concentration increases when the inner cocoon layer is ready, as in larvae spinning a normal cocoon. The larvae which do not spin the outer layer pupate at a lower 20-dihydro-ecdysone level compared with normal larvae. After Giebułtowicz *et al.* (1981), with permission.

in final instar larvae of *Manduca sexta* (Truman, 1978a). It seems, therefore, that it is the appearance of ecdysone that triggers the behaviour associated with the search for a place suitable for metamorphosis, spinning a cocoon or building an appropriate pupation chamber. The decrease in JH concentration, on the other hand, makes the secretion of ecdysone possible (see Chapter 4).

Detailed investigations on the meal moth (*Ephestia kuehniella*) demonstrated that not only the presence of ecdysone, but also its concentration in the tissues, plays a decisive role in the regulation of the behaviour preceding metamorphosis. Depending on environmental conditions such as photoperiod and population density after they emerge from their food, final instar larvae of the meal moth spin a cocoon consisting of two layers—an outer and an inner one (Plate 19)—or build only the inner layer of the cocoon in which they are to pupate (Plate 20). Insects spinning an incomplete cocoon have a much longer wandering period than normal larvae (Giebułtowicz *et al.*, 1980). In larvae with a prolonged wandering period the ecdysone concentration is low and rises very slowly, reaching a maximum during the period when the inner cocoon layer is made. In larvae which produce outer and inner cocoon layers, the dynamics of changes in ecdysteroid content in the body is quite different (Fig. 8.4). Building of the outer cocoon layer stimulates the endocrine system to great activity, and this results in a rapid rise in the ecdysone concentration of the body. It is interesting that in this case the maximum concentration coincides with the spinning of the inner cocoon layer, as in the case of larvae in which the wandering period in prolonged (Giebułtowicz *et al.*, 1981). It seems, therefore, that the behaviour of *Ephestia* larvae associated with the spinning of the outer layer is caused by a drastic increase in 20-hydroxyecdysone concentration in the body. This is confirmed by results of experiments in which large ecdysone doses were injected into younger larvae 1–2 days before the expected time of spinning of the inner cocoon layer. This treatment evoked premature spinning of this layer (see Table 3).

Table 3

Effect of injection of ecdysterone on spinning of inner cocoon layer by young larvae (after Giebułtowicz *et al.*, 1980)

Ecdysterone dose (μg/larva)	No. of larvae	Activity after ecdysterone injection, %		
		spinning of outer layer	spinning of inner layer	no cocoon or incomplete cocoon
1.0	33	3	55	42
0.1	31	29	49	22
control	35	71	23	6

8.3.2. Regulation of behaviour associated with the imaginal moult

Towards the end of imaginal development within the pupa the mechanism regulating eclosion is initiated. At this time, the abdomen of the pupa performs movements that are characteristic for each species. In pupae of *Hyalophora cecropia*, for instance, the premoulting movements of the abdomen start about 1.5 h before the imaginal moult (Truman, 1978a). They consist of rotation of the abdomen and small lateral movements (Fig. 8.5). On the other hand, the prevalent movements during the period of moulting are peristaltic, and these are used by the insect to free itself of the pupal cuticle.

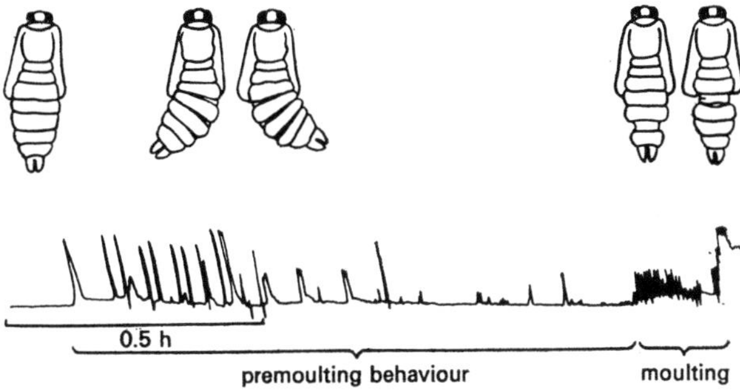

Figure 8.5. Behaviour of *Hyalophora cecropia* pupae before and during imaginal moult. The diagram is a record of abdominal movements of the pupae. Description in text. After Truman (1978), with permission.

As demonstrated by Truman (1973) the activity of pupae before the imaginal moult is stimulated by the eclosion hormone produced in the brain and released into the haemolymph by the *corpora cardiaca*. Further investigations by this author showed that injection of the eclosion hormone into the haemolymph of young pupae produces premature switching on of behaviour appropriate to the moult to the last larval instar. Even the isolated abdomen of such pupae is capable of reacting to this hormone (Truman, 1978b). This is evidence that the nerve centres controlling the action of the corresponding muscle groups are present in the nerve cord. This finding made possible further investigations on the mechanism of action of the eclosion hormone on the central nervous system of *Hyalophora cecropia*. It was found that the central nervous system, removed from the insect's abdomen, reacts by a change in bioelectric activity of particular ganglia (Truman, 1978b). Introduction of the hormone usually switches on, within

15–17 min, the type of bioelectric activity associated with the movements of the abdomen under normal conditions. The mechanism of action of the eclosion hormone on the insect nervous system is discussed in Chapter 4.

8.4. Hormonal regulation of insect locomotor activity

8.4.1. Hormonal control of migration in insects

Certain environmental factors stimulate insects to wander from one place to another. This phenomenon has been examined most thoroughly in the locusts *Locusta gregaria* and *Locusta migratoria migratorioides*. It was found that when these insects are reared singly during the larval stage, the adult forms also lead a solitary life with little motor activity. Rearing in dense groups, on the other hand, leads to the development of individuals that show a tendency to aggregate and are capable of long flights (Johnson, 1969). Examination of the endocrine system of these two forms of locusts revealed wide differences in the morphology of the prothoracic glands (Carlisle and Ellis, 1959). The solitary forms have much larger prothoracic glands than the gregarious forms, in which these organs usually disappear completely. Injection of gregarious forms with a homogenate of prothoracic gland from solitary forms, or implantation of whole glands from solitary insects, greatly reduces their motor activity (Michel, 1972). As mentioned earlier (see p. 180), Haskell and Moorhouse (1963) found that ecdysone inhibits the bioelectric activity of the central nervous system of adult *Schistocerca gregaria* individuals. In view of the results presented here it may be supposed that the gregarious forms of locust have a greatly lowered ecdysone level in the haemolymph. No data confirming this supposition have yet been reported, however.

It seems also that JH participates in the regulation of activity in locusts. Odhiambo (1965) demonstrated that removal of the *corpora allata* in *Schistocerca gregaria* strongly inhibits locomotor activity, and suggests that JH has a stimulating influence on motor activity in this insect. This result, however, has not been confirmed in similar tests with *Locusta m. migratorioides* (Strong, 1968).

The data presented here indicate that both ecdysone and JH may participate in the regulation of insect locomotor activity. However, the evidence requires confirmation in other insects.

Johnson (1969), in his monograph on the regulation of insect migration and dispersion, suggests that the behaviour he describes is due to complete disappearance of ecdysone and the presence of JH in the haemolymph, and that when the JH level rises migration is interrupted and oogenesis begins. This suggestion was confirmed in investigations on *Oncopeltus fasciatus* (Rankin, 1973). Rankin found that implantation of active *corpora allata* or administration of JH analogues to *Oncopeltus fasciatus* females bred under short-day conditions increases the intensity of flight. Injection of a JH analogue into 5-day-old females of this species does not, however, lead to an increase in activity associated with flight, but treatment of younger (2-day-old) females with the analogue stimulates flight. This age-dependent difference disappears completely after removal of the gonads. In gonadectomized insects the JH analogue has the same effect on locomotor activity as in males, that is, it enhances activity. It would seem, therefore, that in *Oncopeltus fasciatus* migration is stimulated by a rise in the JH level in the haemolymph. When sexual maturity is reached, however, migration ceases (see Chapter 7) under the action of this hormone.

A somewhat different mechanism of regulation of migration activities is observed in the Colorado potato beetle. In this species a short photoperiod induces reproductive diapause (see Chapter 6) and migration from the host plant into the soil.

Similar behaviour may be evoked by removal of the *corpora allata* in insects reared under a long photoperiod, which do not normally leave their host plant (de Wilde and de Boer, 1961). Reimplantation of the glands restores the behaviour characteristic of insects reared under a long photoperiod. In the Colorado potato beetle, however, the ovaries do not seem to function in the control of migration in the same way that they do in *Oncopeltus fasciatus* (de Wilde and de Boer, 1969), since females reared under long photoperiod do not change their behaviour after removal of the ovaries. This suggests that in the latter species JH does not influence migratory behaviour by acting on the gonads.

8.4.2. Hormonal regulation of the circadian rhythm of locomotor activity in insects

Like all living organisms, insects have an internal biological clock which is synchronized with diurnal changes in the environment (Cymborowski, 1976). If there are no such changes, the biological clock begins to function according to a circadian rhythm with a period longer or shorter than 24 h. As demonstrated by numerous investigations, this clock controls many types of behaviour, including locomotor activity (Saunders, 1978). This control operates by the neurohormonal pathway (Cymborowski and Brady, 1972). This was demonstrated by means of simple experiments involving the parabiotic coupling of two insects—a rhytmic and a nonrhythmic one—so that they had a common haemolymph circulation (see Plate 11). Parabiosis was achieved by joining the two insects with wax. The legs of the upper insect (rhythmic donor) were removed so that it could not move. It was carried by the recipient—the lower, nonrhythmic insect, the head of which had been removed. Haemolymph flowed from one insect to the other through a short plastic tube. Control parabiosis consisted of a pair of insects with no connection between their haemolymph circulations. After such preparation the insects were transferred to an actograph and their locomotor activity was studied.

It appeared that, owing to the joining of the circulatory systems of the rhythmic and nonrhythmic insects, the latter developed a rhythm in phase with the donor's rhythm. This experiment thus indicates that the biological clock affects the motor activity of insects by means of a chemical factor transported by the haemolymph. Further arduous investigations proved that this factor is produced by the neurosecretory cells of the *pars intercerebralis* of the brain.

The principal observation was that in insects having a distinct diurnal rhythm of locomotor activity there exists a rhythmic accumulation and release of neurosecre-

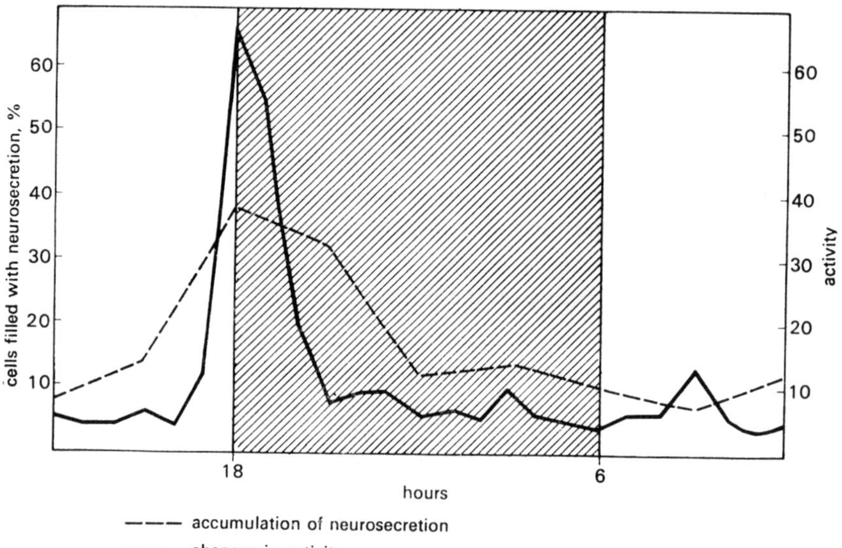

Figure 8.6. Diurnal changes in accumulation and release of neurosecretion in neurosecretory cells of *pars intercerebralis* of the brain of the cricket *Acheta domesticus*. Accumulation of neurosecretion expressed as the percentage of cells filled with this substance in relation to the total number of neurosecretory cells. Changes in diurnal locomotor activity of the insects and the period of darkness are also indicated. After Cymborowski (1970).

tion from these cells (Cymborowski, 1970a, 1973a, b). During the period when the insects exhibit maximal motor activity, i.e. in the first hours of darkness, accumulation of neurosecretion in the brain is also maximal (Fig. 8.6). Release of neurosecretion is associated with a decrease in the intensity of locomotor activity of the insects. The rhythmicity in accumulation and release of neurosecretion is not observed in arrhythmic insects. Such insects can be obtained by rearing under continuous light. This indicates a cause-effect relation between the secretory activity of the neurosecretory cells of the brain and the rhythm of locomotor activity of the insects tested. These results suggest that the neurosecretory material has an inhibitory effect on motor activity. This was confirmed by the results of further experiments involving brain implantation in crickets. These studies showed that, if at least 50 per cent of the medial neurosecretory cells are removed from the brain of a rhythmic insect, it becomes arrhythmic and exhibits a high level of motor activity that is maintained until death (Cymborowski, 1970a). If the brain of a rhythmic donor is implanted into the abdomen of such an insect, the recipient develops a circadian rhythm of motor activity that lasts for 2–3 days after implantation (Fig. 8.7). The rhythm of locomotor activity subsequently disappears and the insect becomes hyperactive, suggesting that the implanted brain survives about three days. The results of this experiment show that even a brain without nerve connections continues to produce and release the neurohormone according to a circadian rhythm, and that this rhythm exerts an inhibitory effect on the locomotor activity of the insect.

Handler and Konopka (1980) demonstrated that in the fruitfly (*Drosophila melanogaster*) transplantation of the biological clock is also possible. By treatment with ethyl methanesulphonate, these authors obtained two mutants differing from each other in their rhythms of spontaneous motor activity. One group consisted of completely arrhythmic insects (per^0) and the other of insects with a spontaneous rhythm of motor activity having a short period (per^s). The period of the rhythm in this group was 18–20 h rather than the 24 h observed in wild-type insects. When the brain of an insect with a short rhythm of locomotor activity was transplanted into an arrhythmic insect, the latter at once became rhythmic. Furthermore, it also exhibited a short rhythm characteristic of the donor's brain. The influence of the implanted brain persisted for 3–5 days. In experiments of this type with crickets the longest period of action of the implanted brain was 2–3 days (Cymborowski, 1970a).

The results presented here have been confirmed in investigations using crickets (Cymborowski, 1981). Two groups of insects differing in the length of the spontaneous locomotor activity period were used. One group consisted of insects

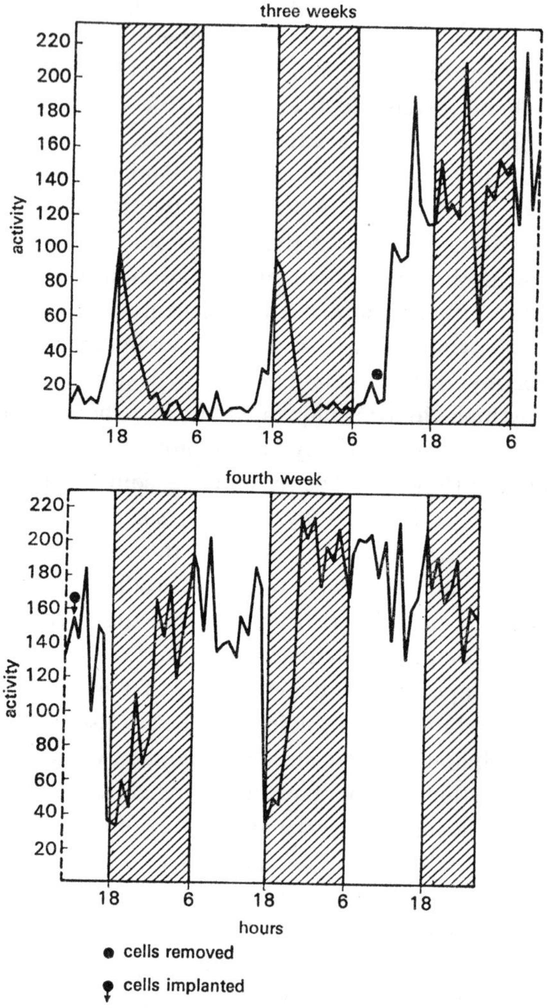

Figure 8.7. Locomotor activity of *Acheta domesticus* in which the neurosecretory cells had been removed from the brain, followed in the fourth subsequent week by implantation of the brain of a cricket showing rhythmic motor activity. Note two periods of depression of locomotor activity after implantation, occurring in periods of darkness (hatched areas). After Cymborowski (1970b).

with a rhythm longer than 24 h, and the second of insects with a shorter rhythm. Implantation of the brain from an insect with a short or long rhythm into an arrhythmic recipient, in which the neurosecretory cells of the *pars intercerebralis* have been destroyed, leads in the recipients to rhythmicity characteristic of the brain of the donor, lasting for 3 days. The implanted brain has inhibitory properties with respect to the motor activity of the recipient which, owing to destruction of its neurosecretory cells, has become hyperactive.

These experiments unequivocally confirm the hypothesis of neurohormonal regulation of locomotor activity in insects. They also indicate that the biological clock of these animals is localized in the *pars intercerebralis*.

8.5. Summary

The regulation of insect behaviour involves the same hormones that regulate development and metamorphosis. The different actions of the same hormones on behaviour consists mainly in the reorganization of the central nervous system during larval moulting and metamorphosis. In the course of these transformations, particular nerve centres are reconstructed; this is manifested by the disappearance of some nerve junctions and the formation of others. This may be most clearly observed during the metamorphosis of the brain and nerve cord ganglia.

The action of hormones on the nervous system also involves a change in sensitivity to external and internal stimuli. Thus, hormones participate in the production of a kind of "mood", so that the insect can react appropriately to these stimuli.

We do not yet know the mechanism of action of hormones on the insect nervous system. They may change the permeability of the nerve cell membranes, influence the rate of neurotransmitter synthesis and transport, or facilitate or block the relay of nerve impulses across synapses, etc.

Insects are an ideal model for investigations of this type, primarily because of the relatively simple structure of their nervous system and the great variety of behaviours characteristic of particular developmental stages. The structure of many insect nerve centres is already known, as are the changes that occur in them during metamorphosis. This may be an excellent starting point for investigations on the mechanism of action of individual hormones, the more so since they can now be synthesized in large quantities.

A particularly promising field of research is the functioning of the biological clock. It is known that the effect of the clock on insect cells and tissues is mediated by neurohormones. The mechanism controlling the rhythmic production and release of these neurohormones, however, remains a mystery.

8.6. References

Barth R. H. Jr. 1968. The comparative physiology of reproductive processes in cockroaches. Part I. Mating behaviour and its endocrine control. *Advan. Reproductive Physiol.*, **3**; 167–207.

Barth R. H., Lester L. J. 1973. Neurohormonal control of sexual behaviour in insects. *Ann. Rev. Entomol.*, **18**; 445–472.

Benz G. 1973. Reversal of spinning behaviour in last instar larvae of *Pieris brassicae* treated with juvenile hormone derivatives. *Experientia*, **29**; 1437–1438.

Birch M. C. 1974. *Pheromones*. North-Holland, American Elsevier, Amsterdam, Oxford, New York.

Bounhiol J. J. 1938. Recherches experimentales sur le déterminisme de la métamorphose chez les Lépidoptères. *Bull. Biol. Fr. Belg.* (*Suppl.*), **24**; 1–199.

Carlisle D. B., Ellis P. E. 1959. La persistance des glandes ventrales céphaliques chez les *Cricquets solitaires*. *C.R. Acad. Sci.*, **249**; 1059–1060.

Ciemior K. E., Sehnal F., Schneiderman H. A. 1979. Moulting growth and survival of *Galleria mellonella* L. (*Lep.*: *Pyralidae*) treated with juvenoids. *Z. angew. Entomol.*, **88**; 414–425.

Cymborowski B. 1970a. Investigations on the neurohormonal factors controlling circadian rhythm of locomotor activity in the house cricket, *Acheta domesticus* L. II. Daily histochemical changes in the neurosecretory cells of the *pars intercerebralis* and suboesophageal ganglion. *Zool. Pol.*, **20**; 127–149.

Cymborowski B. 1970b. Investigations on the neurohormonal factors controlling circadian rhythm of locomotor activity in the house cricket, *Acheta domesticus* L. I. The role of the brain and suboesophageal ganglion. *Zool. Pol.*, **20**; 103–125.

Cymborowski B., Brady J. B. 1972. Insect circadian rhythms transmitted by parabiosis a re-examination. *Nature, New Biol.*, **236**; 221–222.

Cymborowski B. 1973a. Control of the circadian rhythm of locomotor activity of the house cricket. *J. Insect Physiol.*, **19**; 1423–1440.

Cymborowski B. 1973b. Les variations diurnes de l'activité des cellules neurosécrétrices du cerveau du grillon (*Acheta domesticus* L.). *Acrida*, **2**; 299–306.

Cymborowski B. 1976. *Zegary biologiczne* [*Biological clocks*]. PWN, Warszawa.

Cymborowski B., Stolarz G. 1979. The role of juvenile hormone during larval-pupal transformation of *Spodoptera littoralis*: Switchover in the sensitivity of the prothoracic gland to juvenile hormone. *J. Insect. Physiol.*, **25**; 939–942.

Cymborowski B. 1981. Transplantation of circadian pacemaker in house cricket (*Acheta domesticus* L.). *J. interdiscipl. Cycle Res.*, **12**; 133–140.

Dumortier B. 1963. Ethological and physiological study of sound emissions in *Arthropoda*. [In:] *Acoustic Behaviour of Animals*, pp. 583–654.

Engelmann F. 1960. Hormonal control of mating behaviour in an insect. *Experientia*, **16**; 69–70.

Engelmann F., Barth R. H. Jr. 1968. Endocrine control of female receptivity in *Leucophaea maderae* (*Blattaria*). *Ann. Entomol. Soc. Am.*, **61**; 503–505.

Engelmann F. 1970. *The Physiology of Reproduction*, Pergamon Press, Oxford.

Giebułtowicz J. M., Žďárek J., Chróścikowska U. 1980. Cocoon spinning behaviour in *Ephestia kuehniella*; Correlation with endocrine events. *J. Insect Physiol.*, **26**; 459–464.

Giebułtowicz J. M., Delbecque J. P., Cymborowski B. 1981. Correlation of ecdysteroid titers with different premetamorphic behaviours in the larvae of *Ephestia kuehniella* (*Lepidoptera*). [In:] *Regulation of Insect Development and Behaviour*. F. Sehnal, A. Zabża, J. J. Menn, and B. Cymborowski (eds.), pp. 1021–1022. Technical University Press, Wrocław.

Handler A. M., Konopka R. J. 1979. Transplantation of a circadian pacemaker in *Drosophila*, *Nature*, **279**; 236–238.

Haskell P. T., Moorhouse J. E. 1963. A blood-borne factor influencing the activity of the central nervous system of the desert locust. *Nature*, **197**; 56–58.

Hoyle G. 1970. Cellular mechanisms underlying behaviour-Neuroethology. *Adv. Insect Physiol.*, **7**; 349–444.

Hüber F. 1955. Sitz und Bedeutung nervöser Zentren für Instinkt-handlungen beim Männchen von *Gryllus campestris* L. *Z. Tierpsychol.*, **12**; 12–48.

Johnson C. G. 1969. *Migration and dispersal of insects by flight*. Methuen, London.

Loher W. 1966. Die Steuerung sexueller Verhaltensweisen und der Oocytenentwicklung bei *Gomphocerus rufus* L. *Z. vergl. Physiol.*, **53**; 277–316.

Loher W. 1966. Nervöse und hormonale Kontrolle des Sexualverhaltens beim Weibchen der Heuschrecke *Gomphocerus rufus* L. Verhandl. *Deut. Zool. Ges.*, 386–391.

Loher W., Hüber F. 1966. Nervous and endocrine control of sexual behavior in a grasshopper (*Gomphocerus rufus* L., *Acridinae*). *Soc. Exp. Biol. Symp.*, **20**; 381–400.

Lounibos L. P. 1974. *The Cocoon Spinning Behaviour of Saturniid Silkmoths*. Ph. D. Thesis, Harvard University.

Markl H. 1974. Insect behavior: Function and mechanisms. [In:] *The Physiology of Insecta*. III. M. Rockstein (ed.). Acad. Press, New York, London.

Michel R. 1972. Étude expérimentale de l'influence des glandes prothoraciques sur l'activité de vol du Criquet Pélerrin *Schistocerca gregaria*. *Gen. Comp. Endocrinol.*, **19**; 96–101.

Odhiambo T. R. 1965. Metabolic effects of the *corpus allatum* hormone in the desert locust, *Schistocerca gregaria*. *Nature*, **207**; 1314–1315.

Ozbas S., Hodgson E. S. 1958. Action of insect neurosecretion upon central nervous system *in vitro* and upon behavior. *Proc. Nat. Acad. Sci. USA*, **44**; 825–830.

Piepho H. 1950. Hormonale Grundlagen der Spinntätigkeit beu Schmetterlingsraupen. *Z. Tierphysiol.*, **7**; 424–434.

Rankin M. A. 1973. The hormonal control of flight in the milkweed bug, *Oncopeltus fasciatus*. [In:] Experimental Analysis of Insect Behaviour. L. B. Browne. (ed.). Springer-Verlag, Berlin.

Riddiford L. M. 1970. Antennal proteins of saturniid moths: their possible role in olfaction. *J. Insect Physiol.*, **16**; 653–660.

Riddiford L. M., Williams C. M. 1971. Role of the *corpora cardiaca* in the behaviour of saturniid moths. I. Release of sex pheromone. *Biol. Bull.*, **140**; 1–7.

Riddiford L. M. 1972. Juvenile hormone in relation to the larval-pupal transformation of the *cecropia* silkworm. *Biol. Bull.*, **142**; 310–325.

Röller H., Piepho H., Molz J. 1963. Zum Problem der Hormonabhängigkeit des Paarungsverhaltens bei Insekten. Untersuchungen an *Galleria mellonella* (L.). *J. Insect Physiol.*, **9**; 187–194.

Safranek L., Cymborowski B., Williams C. M. 1980. Effects of juvenile hormone on ecdysone-dependent development in the tobacco hornworm, *Manduca sexta*, *Biol. Bull.*, **158**; 248–256.

Saunders D. S. 1976. *Insect Clocks*. Pergamon Press, Oxford, New York, Toronto, Sydney, Paris, Frankfurt.

Sieber R., Benz G. 1978. The influence of juvenile hormone on the feeding behaviour of last instar larvae of the codling moth, *Laspeyresia pomonella* (*Lep*., *Tortricidae*), reared under different photoperiods. *Experientia*, **34**; 1647–1650.

Sláma K., Romaniuk M., Šorm F. 1974. *Insect Hormones and Bioanalogues*. Springer-Verlag, Wien, New York.

Strong L. 1968. The effect of enforced locomotor activity on lipid content in allatectomized males of *Locusta migratoria migratorioides*. *J. Exp. Biol*., **48**; 525–530.

Truman J. W. 1971. Physiology of insect ecdysis. I. The eclosion behavior of saturniid moths and its hormonal release. *J. Exp. Biol*., **54**; 805–814.

Truman J. W., Sokolove P. G. 1972. Silkmoth eclosion: hormonal triggering of a centrally programmed pattern of behavior. *Science*, **175**; 1491–1493.

Truman J. W. 1973. How do moths "turn on"? A study of the action of hormones on the nervous system. *Amer. Sci*., **61**; 700–706.

Truman J. W., Riddiford L. M. 1974. Hormonal mechanism underlying insect behaviour. *Adv. Insect Physiol*., **10**; 297–362.

Truman J. W. 1976. Development and hormonal release of adult behavior patterns in silkmoths. *J. Comp. Physiol*., **107**; 39–48.

Truman J. W., Riddiford L. M. 1977. Invertebrate system for the study of hormonal effects on behavior. [In:] *Vitamins and Hormones*, pp. 283–315. Academic Press, New York, San Francisco, London.

Truman J. W. 1978a. Hormonal control of invertebrate behavior. *Horm. Behav*., **10**; 214–234.

Truman J. W. 1978b. Hormonal release of stereotyped motor programs from the isolated nervous system of the *Cecropia* silkmoth. *J. Exp. Biol*., **74**; 151–174.

Wilde J., de, de Boer J. A. 1961. Physiology of diapause in the adult Colorado beetle. II. Diapause as a case of pseudoallatectomy. *J. Insect Physiol*., **6**; 152–161.

Wilde J., de, de Boer J. A. 1969. Hormonal and nervous pathways in photoperiodic induction of diapause in *Leptinotarsa decemlineata*. *J. Insect Physiol*., **15**; 661–675.

Wilson E. O., Bossert W. H. 1963. Chemical communication among animals. *Rec. Prog. Horm. Res*., **19**; 673–716.

Wilson E. O. 1971. *The Insect Societies*. The Belknap Press of Harvard University Press, Cambridge, Massachusetts.

Ždárek J., Sláma K. 1968. Mating activity in adultoids and supernummerrary larvae induced by agents with high juvenile hormone activity. *J. Insect Physiol*., **14**; 563–567.

Ždárek J. 1970. Mating behavior in the bug, *Pyrrhocoris apterus* L. (*Heteroptera*): Ontogeny and its environmental control. *Behaviour*, **37**; 252–267.

Ždárek J. 1971. Hormonal control of mating behavior in *Pyrrhocoris apterus* L. (*Heteroptera*). *Proceedings of the International Symposium in Insect Endocrines, Brno* 1966. 51–61.

Ždárek J., Kontew Ch. 1975. Some ethological aspects of reproduction in *Eurygaster integriceps* (*Heteroptera Scuttelleridae*). *Acta entomol. bohemoslov*., **72**; 239–248.

Ždárek J. 1981. The role of hormones in the control of premetamorphic behaviour in insects. [In:] *Regulation of Insect Development and Behaviour*. F. Sehnal, A. Zabża, J. J. Menn and B. Cymborowski (eds.), pp. 993–1010. Technical University Press, Wrocław.

Chapter 9

Insect hormones as insecticides

9.1. General characteristics of third-generation insecticides

The idea of using hormones in insect control was advanced for the first time by Williams (1956). This prompted many investigators to an intensive search for hormone-mimetic substances resembling natural insect hormones. Within a short time an enormous amount of data concerning the influence of substances regulating insect development and metamorphosis was amassed (Sláma *et al.*, 1974; Laskowska-Bożek and Zielińska, 1978). Williams (1967) named these substances third-generation pesticides. Nowadyas they are commonly known as insect growth regulators (IGR) (Staal, 1975; Oberlander *et al.*, 1978).

These compounds differ widely from the commonly used insecticides. Their principal effect is to disturb development and metamorphosis, and their action is therefore much slower than that of conventional insecticides, which act mainly on the nervous system and kill the insect almost immediately. Since these compounds disturb insect metamorphosis, they are highly specific, and do not affect other groups of invertebrates or vertebrates (Staal, 1975).

The effectiveness of IGR action depends largely on the developmental stage of the insect and on its physiological state; it is greatest when the compound enters the insect's body during the period when, under physiological conditions, natural hormones are not present. In the case of juvenile hormone such a period occurs, as already mentioned, in the second half of the last larval instar (see Chapter 4). It is therefore in this period that insects are most susceptible to the action of exogenous JH or JH analogues. These substances cause irreversible physiological and morphological changes, resulting in the death of the insect. There are also periods when ecdysone does not occur in the insect body. Administration of exogenous ecdysone or an analogue may accelerate or retard development, leading to physiological as well as morphological changes, and to death.

Substances are also now known that inhibit the synthesis or action of these hormones. They are therefore antihormones, producing effects opposite to those of JH and ecdysone.

Beside these four main types of IGR, there are also compounds that disturb cuticle synthesis. The best known compound of this type is Dimilin (TH6040). Later in this chapter the mechanisms of action of the different types of IGR will be discussed with particular reference to substances imitating the action of JH (juvenoids), which may be of practical importance in insect control.

9.2. Juvenoids

The greatest hopes in the application of compounds imitating the action of JH were associated with the discovery by Sláma and Williams (1965, 1966) of the "paper factor" disturbing development in *Pyrrhocoris apterus*. When the insects were reared in contact with paper made from balsam fir, the population died out owing to the development of sterile individuals, additional larval stages and forms transitional between larvae and imagines. Soon Bowers *et al.* (1976) isolated the compound producing this effect, and named it juvabion (Fig. 9.1).

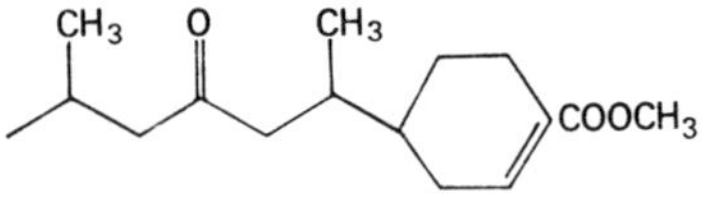

Figure 9.1. Structural formula of juvabion.

This substance and dehydrojuvabion, isolated by a group of investigators in Prague from a species of fir growing in Slovakia, exhibit an unusually high biological activity. It should also be stressed that these substances act in this way only in species of the family *Pyrrhocoridae*. Unfortunately, however, these insects for the most part are not pests. This discovery led to the production within about 12 years of more than 5000 compounds imitating the action of juvenile hormone, which were termed juvenoids. Most of them have a highly specific action (Sláma *et al.*, 1974). Like JH they are very unstable and decompose under UV radiation, so that they cannot accumulate in the natural environment, as DDT has. The instability of these compounds leads to many practical difficulties in their application. They have to be used in combination with protective substances (Quistad *et al.*, 1974; Sláma *et al.*, 1974).

9.2.1. Methods of application of juvenoids

Problems associated with juvenoid application have been discussed in numerous publications (Wigglesworth, 1963, 1969; Jarolim et al., 1969; Sláma

1971; Sláma *et al.*, 1974; Staal, 1975; Oberlander *et al.*, 1978). In laboratory investigations on juvenoids, organic solvents such as ethanol, acetone, octane etc. are used, mainly because they have no effect on the development and metamorphosis of insects. Dissolved juvenoids are usually injected into the body cavity of the insects. This is, however, a highly time-consuming and frequently ineffective method, because when introduced into the haemolymph these compounds are very quickly excreted from the body (Němec and Jarolim, 1980). The common methods for the application of juvenoids are therefore: (1) surface contact, (2) fumigation, and (3) in the food. In the contact method, the dissolved juvenoid is placed on the body surface or on a filter paper disc on which the insect is restrained. This allows gradual and prolonged penetration of the substance into the haemolymph. This increases the probability that the juvenoid penetrates when the insects are most susceptible to it. To reduce the rate of penetration of JHA into the body cavity still more, oil or even wax are frequently used as the solvent in *Galleria mellonella* bioassays (see Chapter 3).

Fumigation with juvenoid vapours may find wide application when access to the insects is difficult, as in the case of insects living in grain stores. Pupae in cocoons can also be treated by fumigation (Metwally and Sehnal, 1973). Application in the food may facilitate the penetration of JHA to insects living on stored products. In this case, however, the juvenoid acts mainly on the feeding larval stages and some adult forms (Williams and Amos, 1974; Oberlander *et al.*, 1978).

9.2.2. Main types of juvenoids

Natural forms of JH have never been seriously considered from the viewpoint of practical application. The possibility of synthesizing their bioanalogues led to the production of compounds with much higher activity, stability and selectiveness of action, and a low cost of production. Some of these compounds are presented in Figure 9.2. Details of these and many other juvenoids appear in numerous publications (Staal, 1975; Menn and Pallos, 1975; Sehnal, 1976) and will not be given here.

Juvenoids can be divided into a number of groups. Firstly there are the compounds with a main carbon skeleton of the farnesane type, similar in structure to the natural hormone and differing only in the substituents at C-3, C-7 and C-11 and in the substituent in the ester group. This group is most active in lepidopteran larvae. The second group, which is very numerous, consists of sesquiterpenoid derivatives and includes derivatives of 2,3-dodecadienic acid, which are

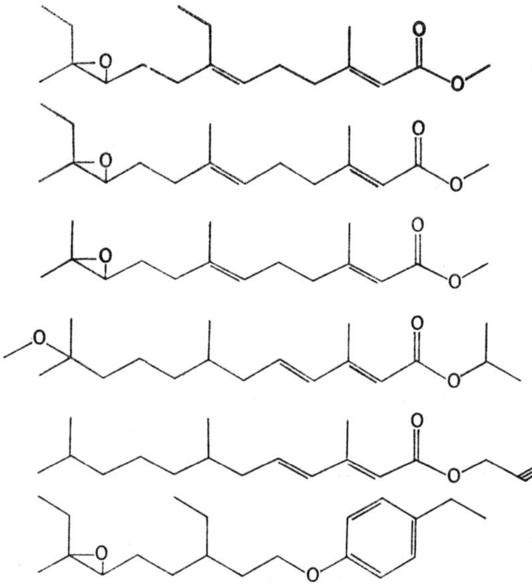

Figure 9.2. Chemical structure of natural juvenile hormones (JHI, JHII, JHIII) and of some of the most active bioanalogues of practical importance.

present in Altosid, a preparation used for mosquito control. This substance is also active against other dipteran insects, such as house flies (Henrick *et al.*, 1973).

To this group of juvenoids also belong compounds exhibiting high activity towards beetles (Sláma *et al.*, 1974).

The third group of juvenoids consists of aromatic esters, thioethers and amines with a geranyl or farnesane radical. Epophenane is an example of this group of compounds (Fig. 9.2), and has high stability in field conditions (Hangarter *et al.*, 1976).

Another group consists of peptide juvenoids (Poduška *et al.*, 1971). Some of these exhibit extremely high biological activity, particularly towards the *Pyrrhocoridae*, although this family does not include any dangerous pests.

Sláma and Romaňuk (1976) elaborated a new group of juvenoids which they called juvenogens. They are rather complicated esters which include in their composition active juvenoids and long-chain fatty acids. On entering the haemolymph this complex is hydrolyzed in a manner similar to the normal inactivation of juvenoids. However, an active juvenoid is formed from the juvenogen. Instead of inactivating the hormone, therefore, the insect begins to produce it. Before the eventual introduction of these compounds for insect control, further intensive investigations are necessary, both as regards the possibility of synthesizing them and their influence on insect physiological processes. A great advantage of juvenogens is their higher stability under natural environment conditions compared with juvenoids themselves. They readily penetrate through the insect cuticle, a very convenient property in practical application.

In addition to the above groups of juvenoids others are known, such as terpenoid derivatives with a geraniol skeleton. However, on account of their low biological activity there are no prospects of their application in insect control. Many of them are discussed in a separate monograph (Menn and Beroza, 1972).

9.2.3. Morphogenetic action of juvenoids

Juvenoids affect practically all development stages of insects (Fig. 9.3). The most sensitive stage, however, is the egg, in which embryonic development

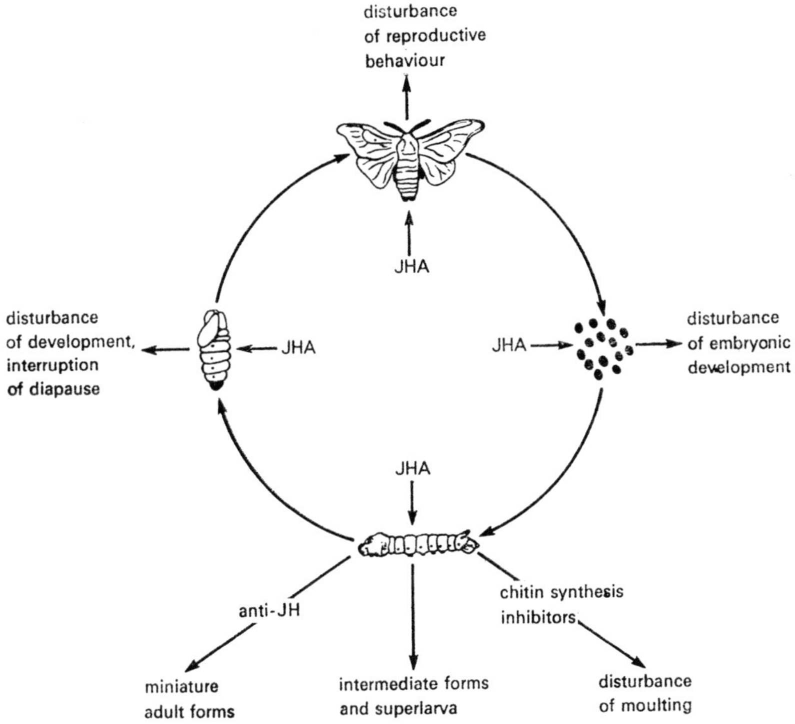

Figure 9.3. Possible actions of juvenoids at various stages of the insect developmental cycle. Further explanation in text.

takes place, and the last larval instar, during which the processes associated with metamorphosis begin. This is no doubt due to the fact that in these stages of development the endogenous JH level is very low. In the developing embryo, for instance, the *corpora allata* do not function, while in the last larval stage their activity is switched off before metamorphosis (see Chapter 4). Larval stages are also very sensitive to inhibitors of the processes involved in chitin synthesis and to antihormones, because of their frequent moults.

9.2.3.1. Influence on embryogenesis

Juvenoids affect embryogenesis mainly by disturbing it, so that the larvae do not hatch (Sehnal, 1976). In all cases that have been studied, this effect was obtained only when juvenoids were applied to freshly laid eggs. Very high juvenoid concentrations are necessary to obtain the same effect at later stages. Application of JHA after blastokinesis usually merely retards further ontogenetic development, owing to the occurrence of additional larval moults (Masner *et al.*, 1968; Riddiford, 1970, 1972; Hunt and Shappirio, 1973). It seems, therefore, that it is during embryonic development that the programming of the duration of the entire period of ontogenetic development takes place. As claimed by Riddiford (1972), the amount of juvenoid needed to change this programme is three times higher than that required to prevent the hatching of larvae from eggs. This is probably caused by the difference in endogenous JH concentration between freshly laid eggs and developing embryos.

9.2.3.2. Influence on metamorphosis

Treatment of final-instar larvae or pupae with juvenoids disturbs the morphogenetic processes leading to the development of the imago.

These disturbances consist in the appearance of the reversible and irreversible changes (Sehnal, 1976) shown in Table 4.

As regards the reversible changes, the insects develop into imagos, but metamorphosis is considerably delayed. The individuals are usually capable of reproduction and their progeny are normal. Superlarvae of *Galleria mellonella* formed as the result of juvenoid action, for instance, are much larger than normal last-instar larvae (Plate 21), but do not differ from the latter in morphology. In the case of irreversible changes, however, the individuals die because of abnormal morphological development (Plate 22). The extent of these morphological changes depends on many factors, such as the species involved, time of application

Table 4
Influence of juvenoids on metamorphosis

Reversible changes	Irreversible changes
Protraction of the last larval stage:	The insects develop into:
a. continuation of feeding (growth of larva)	a. imperfect superlarvae
b. feeding ceases (typical larval diapause)	b. transitional forms between preceding and following stage of development
Additional larval stages	c. pupae and adult forms with changes in internal organs
a. continuation of feeding and growth	
b. limited feeding	
c. feeding ceases (diapause)	

of the juvenoid, the way in which it is administered, and the type of juvenoid. Administration of a suitable dose at the very beginning of the last instar usually leads to development of additional larval instars, whereas treatment during the second half of this stage produces irreversible morphological changes. This is probably due to the fact that during the period when the endogenous JH level in the haemolymph is gradually falling, the metamorphosis programme is gradually switched on in the cells of different tissues. The initiation of this programme in some cells and tissues cannot be reversed by juvenoids. On the other hand, juvenoids are capable of switching on this programme in other cells and tissues, and this produces a desynchronization effect with respect to metamorphosis in different organs and tissues. Thus, there is a critical period in larval development when juvenoids exert the strongest effect (Sehnal and Meyer, 1968).

From the practical point of view, it is the production of irreversible morphological changes in the field that is the most desirable effect of juvenoid application. Development of additional larval stages merely increases the damage caused by the insects as a result of prolonged feeding. This is a serious inconvenience in the use of juvenoids for insect control. Their application requires accurate identification of the developmental stages of the pests in the field. This is not a simple matter, and the persons responsible for the work have to be specially trained. Another drawback of juvenoids is that the period of sensitivity to their action is relatively short, so that the field treatment has to be carried out rapidly. Another disadvantage is the instability of juvenoids under field conditions, mentioned previously.

In spite of these negative properties of juvenoids, numerous attempts have been made to apply them in practice (Staal, 1975; Sehnal, 1985; Rowlands and Edwards, 1986).

9.2.4. Juvenoids in insect control

The first juvenoid, methoprene—commercial name Altosid (see Fig. 9.2)—was released for general use in the United States as early as 1975, and is now commonly used for the control of mosquitoes (Djerassi, 1975). Addition of this compound, even in trace amounts, to stagnant water completely prevents the development of mosquito larvae. Its use is particularly important in tropical countries and those suffering from frequent floods, where mosquitoes find favourable conditions for development.

Takahashi and Miura (1975) did not find any negative effects of this compound on plant and animal groups (e.g. fish), and on humans. The degree of safety of its use in the control of different fly species, especially *Haematobia irritans*, the larvae of which develop in cattle faeces, may be measured by the fact that its use is permitted in the USA. In the control of *Haematobia irritans*, is given in drinking water or added to the food at the rate of 0.7 mg/kg body weight per 24 h. It is also administered in mineral blocks, which are readily licked by animals on account of their salt content (Chamberlain, 1975). The use of this compound reduced the fly population by about 99 per cent. This is an achievement worth stressing in view of the losses in milk and meat production caused by these insects.

Precise analysis of milk and meat from animals receiving methoprene did not show even the smallest traces of the compound.

Encouraging results have also been obtained in the use of methoprene to control houseflies in large chicken farms (Morgan *et al.*, 1975). In this case too the compound could not be detected in chicken meat.

In the control of forest pests, epophenane, another juvenoid that shows high biological activity (see Fig. 9.2), seems very promising. Field experiments appear to have been successful (Hangartner *et al.*, 1976; Vogel *et al.*, 1976).

Juvenoids may also find application in the control of social insects such as pharaoh's ants, which have infested numerous houses in Warsaw, and termites (Cupp and O'Neal, 1973; Edwards, 1975). The application of these compounds to social insects is aided by the worker insects, which disseminate them throughout the nest.

Practical application of juvenoids (especially methoprene) is of interest in the rearing of the silkworm (*Bombyx mori*). Methoprene has been used in Japan since 1976 to increase the weight of the cocoons. Addition of methoprene to the insect' food at a rate as low as 10 ppm increases silk production by 10–20 per cent without affecting quality (Ozawa *et al.*, 1973).

To sum up the results obtained to date, it may be said that juvenoids have the following principal advantages:

1. the period required for complete degradation is short,
2. their activity is specific—they act on only one order or species of insects, and not on others; some act only on certain developmental stages,
3. no side effects have been noted; they can be applied in minimal concentrations,
4. they are easy and cheap to produce on an industrial scale.

9.3. Ecdysoids

A large number of bioanalogues of the second important insect hormone, ecdysone, are available (Piechowska *et al.*, 1970; Sláma *et al.*, 1974). Since their

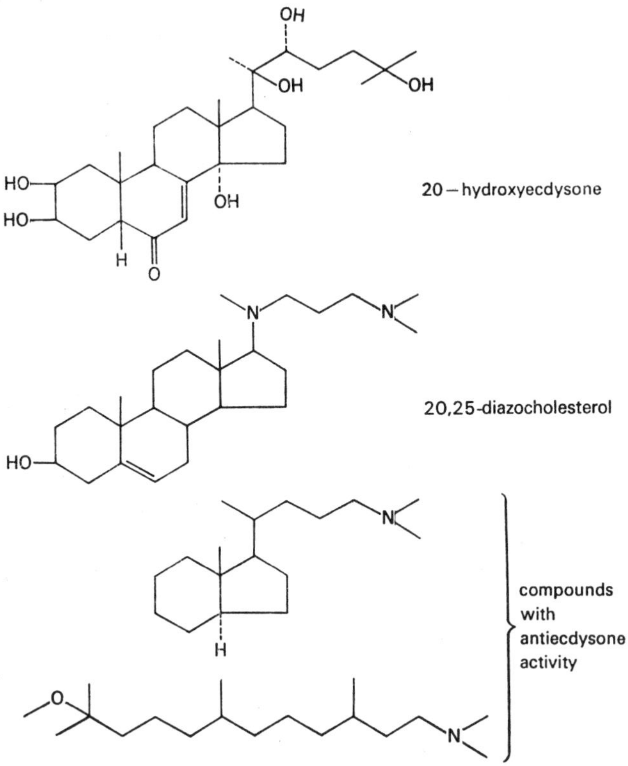

Figure 9.4. Chemical structure of 20-hydroxyecdysone, 20,25-diazocholesterol and compounds with a relatively simple structure showing antiecdysone activity. After Staal (1977), with permission.

steroid structures are extremely complicated (see Fig. 9.4), the synthesis of analogues (ecdysoids) of this type is not economically profitable. However, plants are a very rich source of phytoecdysones, which are rather easy to isolate and purify (Sláma, 1980). They are hydrophilic rather than lipophilic, and their use in insect control is therefore very difficult. They cannot penetrate the wax layer that covers the insect cuticle and have to be introduced into the body with food. Moreover, laboratory investigations have shown that high ecdysoid doses are necessary to produce changes in insect metamorphosis, which of course limits the possibility of applying these compounds in practice (Staal, 1977). The effects of ecdysoids on other animal groups are not clear. Experiments have shown that they cause abnormalities in the growth of mammalian cells *in vitro* (Burdette and Richards, 1961), and produce toxic biochemical changes in the liver (Otaka *et al.*, 1969). On the other hand, there are also suggestions that they may exert a favourable influence in human neurological diseases (Hikino and Takemoto, 1972).

So far the only practical application of ecdysoids is in silkworm breeding in Japan. Spraying the larvae with ecdysoid in the rearing laboratory results in synchronization of the spinning of cocoons, and this greatly lowers the cost of silk production.

9.4. Antihormones

9.4.1. Antijuvenile compounds

Since the juvenile hormone is indispensable for growth and development of insects, removal of its source (allatectomy) should cause inhibition of development. Experiments of this type have been performed in many laboratories, giving unequivocal results. After removal of the *corpora allata*, larvae cease feeding almost immediately and begin to prepare for metamorphosis. The younger the larvae when the *CA* are removed, the smaller are the insects that develop from them (Fig. 9.5), and the smaller should then be the economic losses. A natural substance was sought which would block the function of the *CA*, thus causing premature metamorphosis. Soon Bowers (1976) obtained an extract with antijuvenile activity from the common plant *Ageratum houstonianum*. Analysis of this extract revealed two active compounds belonging to the group of ageratochromenes, which were designated precocenes. Precocene II (Fig. 9.6) proved highly active, especially towards *Oncopeltus fasciatus* larvae (Bowers *et al.*, 1976). Application of this substance to larvae of the second developmental stage led to

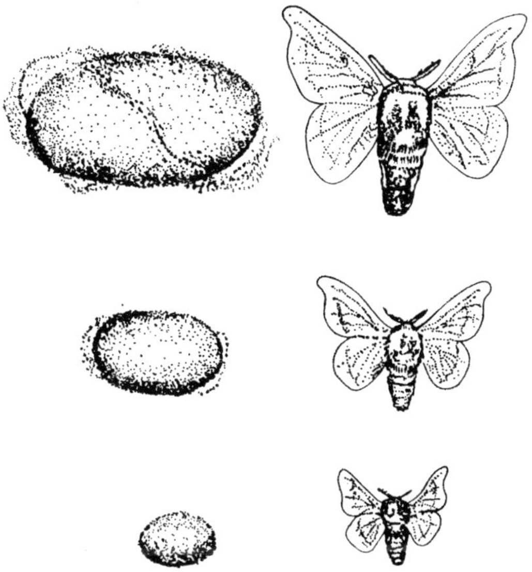

Figure 9.5. Effect of removal of *corpora allata* at various larval stages in *Bombyx mori*. The earlier this is done the smaller are the adult forms that develop. Top: cocoon and control insect. Middle and bottom: after removal of *corpora allata*.

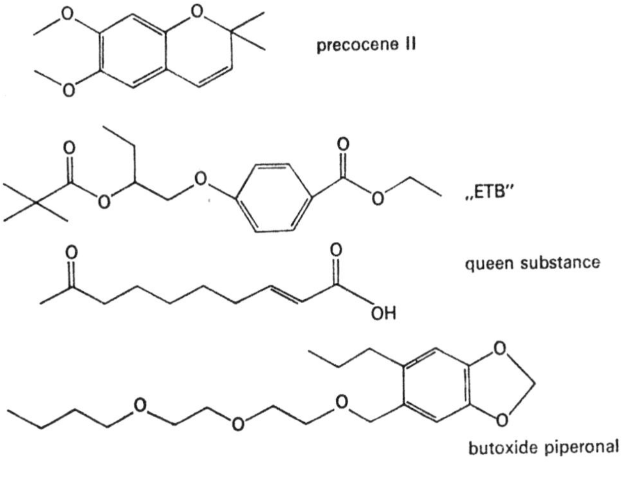

Figure 9.6. Chemical structure of some compounds showing antijuvenile hormone activity. The antijuvenile compound "ETB" is active in *Manduca sexta* larvae. After Staal (1977), with permission.

formation of a normal third instar, and subsequently to a miniature imago (Fig. 9.7). Thus, two successive larval stages were omitted. A similar precocious metamorphosis was obtained in the migrating locust (Nemec *et al.*, 1978; Pedersen, 1978).

The mechanism of action of precocenes has not yet been fully elucidated. It seems that they may act as antifeedants (Sláma, 1978). Essentially, however, they evoke premature atrophy of the *corpora allata* (Unnithan *et al.*, 1977; Pener *et al.*, 1978) and also perhaps of the prothoracic glands (Masner *et al.*, 1976). Of course, blocking of *CA* function by precocenes causes simultaneous underdevelopment of the gonads because, as already mentioned (Chapter 5), JH functions as a gonadotropic hormone.

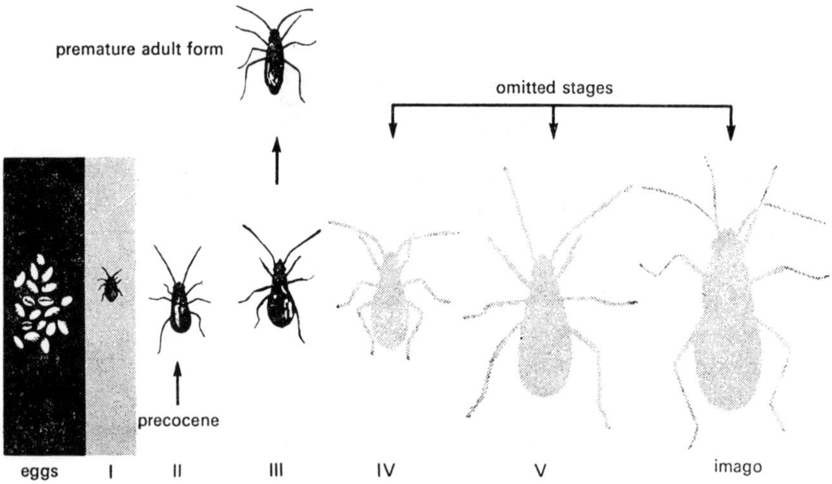

Figure 9.7. Influence of precocene on development of *Oncopeltus fasciatus*. Administration of this compound in the second larval instar causes moulting to third instar, from which a miniature adult insect emerges. Thus, two larval instars are omitted. After Bowers (1977), with permission.

Further intensive studies demonstrated that precocenes act only on insects with incomplete development, but have no noticeable influence on holometabolous species. "ETB" compounds (see Fig. 9.6), however, show some antijuvenile action, which is strongest in *Manduca sexta* larvae (Staal, 1977).

Other compounds shown in Figure 9.6 show antijuvenile activity in some species of *Lepidoptera*.

9.4.2. Antiecdysteroids

Hora *et al.* (1966) found that some steroids similar in structure to ecdysone inhibit the hardening and sclerotization of the cuticle after moulting in *Pyrrhocoris apterus*. Implantation of several crystals (20–50 μg) of such a compound into the body cavity of freshly moulted larvae prevents the cuticle hardening for about 12 h. This result has been confirmed by other investigators (Robbins *et al.*, 1968), and many substances of this type, acting antagonistically to ecdysone, are now known (Sláma *et al.*, 1974). Application of these compounds in practice is, however, severely limited because they are active only after injection into the haemolymph.

The same is true for azosterols and simpler compounds containing nitrogen (see Fig. 9.4), which exhibit antiecdysteroid properties, but only when they penetrate into the haemolymph (Staal, 1977). Their mechanism of action is different from that of the compounds mentioned above. When ingested with vegetable food, they inhibit the metabolism of sitosterol to cholesterol, a precursor of ecdysone.

Further research may lead to the detection of many other compounds counteracting the function of ecdysone; the synthesis of this hormone in the insect is very complicated, giving the possibility of many kinds of interference.

9.5. Summary

Too many hopes seem to have been attached to the practical application of hormonomimetic compounds in insect control. Of course, they cannot be universal agents in this fight. Their application, like that of other insecticides, has both negative and positive effects, the most important of which are their high specificity of action and failure to accumulate in the environment.

It was at first believed that since the juvenile hormones are necessary for complete development and metamorphosis, their analogues will not induce an insect's resistance. This, however, is not true. Certain strains of houseflies are known to be resistant to juvenoids (Cerf and Georghiou, 1972) and mosquitoes (Brown and Brown, 1974). These results, however, were obtained in laboratories and are not necessarily applicable in the field. No data on this aspect are available as yet. No doubt juvenoids, in combination with other insecticides, will in the future play an important role in the integrative method of control of insect population density.

9.6. References

Bowers W. S., Fales H. M., Thompson M. J., Uebel E. C. 1966. Juvenile hormone: Identification of an active compound from balsam fir. *Science*, **154**; 1020–1021.

Bowers W. S. 1976. Discovery of insect antiallatotropins. [In:] *The Juvenile Hormones*. L. J. Gilbert (ed.), pp. 394–408. Plenum Press, New York–London.

Bowers W. S., Ohta T., Cleere J. S., Marsella P. A. 1976. Discovery of insect anti-juvenile hormones in plants. *Science*, **193**; 542–547.

Brown T. M., Brown A. W. A. 1974. Experimental induction of resistance to a juvenile hormone mimic. *J. Econ. Entomol.*, **67**; 799–801.

Burdette W. J., Richards R. C. 1961. Alternation of growth of mammalian cells *in vitro* by ecdysone extract. *Nature*, **189**; 666–668.

Cerf D. C., Georghiou G. P. 1972. Evidence of cross-resistance to a juvenile hormone analogue in some insecticide resistant houseflies. *Nature*, **239**; 401–402.

Chamberlain W. F. 1975. Insect growth regulating agents for control of arthropods of medical and veterinary importance. *J. Med. Entomol.*, **12**; 394–400.

Cupp E. W., O'Neal J. 1973. The morphogenetic effect of two juvenile hormone analogues on larvae of important fire ants. *Environ. Entomol.*, **2**; 191–194.

Djerassi G. 1975. First products, marketing and manufacturing plants. Zöecon Corporation. Ann. Report for the year, ended December 1974.

Edwards J. P. 1975. The effect of a juvenile hormone analogue on laboratory colonies of pharaoh's ant. *Monomorium pharaonis* (L.) (*Hymenoptera, Formicidae*). *Bull. Entomol. Res.*, **65**; 75–80.

Hangartner W. W., Suchy M., Wipf H-K., Zurflueh R. C. 1976. Synthesis and laboratory and field evaluation of a new, highly active and stable insect growth regulator. *J. Agric. Food Chem.*, **24**; 169–175.

Henrick C. A., Staal G. B., Siddal J. B. 1973. Alkyl 3,7,11-trimethyl-2,4-dodecadienoates, a new class of potent insect growth regulators with juvenile hormone activity. *J. Agr. Food Chem.*, **21**; 354–359.

Hikino H., Takemoto T. 1972. Arthropod moulting hormones from plants *Achyranthes* and *Cyathula*. *Naturwiss.*, **59**; 91–98.

Hora J., Lábler L., Kasal A., Černý Ž., Šorm F., Sláma K. 1966. Moulting deficiencies produced by some sterol derivatives in an insect *Pyrrhocoris apterus* L., *Steroids*, **8**; 887–914.

Hunt L. M., Shappirio D. G. 1973. Larval development following juvenile hormone treatment of eggs of the lesser milkweed bug, *Lygaeus kalmii*. *J. Insect Physiol.*, **19**; 2129–-2134.

Jarolim V., Hejno K., Sehnal F., Šorm F. 1969. Natural and synthetic materials with insect hormone activity. Juvenile activity of the farnesane-type compounds on *Galleria mellonella*. *Life Sci.*, **8**; 831–841.

Laskowska-Bożek H., Zielińska Z. M. 1978. Analogi hormonów juwenilnych jako insektycydy trzeciej generacji. Fizjologiczne podstawy [Juvenile hormone analogues as third generation insecticides. Physiological bases]. *Kosmos*, **4**; 367–382.

Masner P., Sláma K., Landa V. 1968. Sexually spread insect sterility induced by the analogues of juvenile hormone. *Nature*, **219**; 39–396.

Masner P., Vogel W., Graf O., Bowers W. S. 1976. Mode of action of insect growth regulators with juvenile hormone and antiallatotropical activity. *XV Int. Congr. Entomol., Washington D. C. USA. Aug.*, 19–27.

Menn J. J., Beroza M. (eds.) 1972. *Insect Juvenile Hormones. Chemistry and Action.* Academic Press, New York, London.

Menn J. J., Pallos F. M. 1975. Development of morphogenetic agents in insect control. *Environ. Lett.*, **8**; 71–88.

Metwally M. M., Sehnal F. 1973. Effect of juvenile hormone analogues on the metamorphosis of beetles *Trogoderma granarium* (*Dermestidae*) and *Caryedon gonagra* (*Bruchidea*). *Biol. Bull.*, **144**; 368–382.

Morgan P. B., LaBrecque G. C., Weidhaas D. E., Benton A. 1975. The effect of metoprene, an insect growth regulator, on *Musca domestica. Can. Entomol.*, **107**; 413–417.

Némec V., Chen T. T., Wyatt G. R. 1978. Precocious adult locust, *Locusta migratoria migratorioides*, induced by precocene. *Acta entomol. bohemoslov.*, **75**; 285–286.

Němec V., Jarolim V. 1980. Excretion of some juvenoids by *Locusta migratoria* larvae. *Acta entomol. bohemoslov.*, **77**; 76–81.

Oberlander H., Nickle D., Silhacek D. L., Hagstrum D. W. 1978. *Proc. Symp. on Prevention and Control of insects in Stored-Food Prod., Manhattan, K.s., April*, **9–13**; 247–263.

Otaka T., Okuji S., Uchiyama M. 1969. Stimulation of protein synthesis in mouse liver by ecdysterone. *Chem. Pharm. Bull.*, **17**; 75–81.

Ozawa J., Mori K., Matsui M. 1973. Synthesis and biological activity of analogues of the *Cecropia* juvenile hormone with different alkyl substituents at C-7 and C-11. *Agr. Biol. Chem.*, **37**; 2373–2378.

Pedersen L-E. K. 1978. Effects of anti-juvenile hormone (Precocene I) on development of *Locusta migratoria* L. *Gen. Comp. Endocrinol.*, **36**; 502–509.

Pener M. P., Orshan L., de Wilde J. 1978. Precocene II causes atrophy of *corpora allata* in *Locusta migratoria. Nature*, **272**; 350–353.

Piechowska H., Sienkiewicz Z., Bielińska M. 1970. Hormony metamorfozy owadów [Hormones of insect metamorphosis]. *Post. Biochem.*, **16**; 449–481.

Poduška V., Šorm F., Sláma K. 1971. Natural and synthetic materials with insect hormone activity. 9. Structure—juvenile hormone activity relationship in simple peptides. *Z. Naturforsch.*, **26b**; 719–722.

Quistad G. B., Staiger L. E., Schooley D. 1974. Environmental degradation of the insect growth regulator methoprene (isopropyl) 2E, 4E-(-11-methoxy-3,7,11-trimethyl-2,4-dodecadienoate). I. Metabolism by alfalfa and rice. *J. Agr. Food Chem.*, **22**; 582–589.

Riddiford L. M. 1970. Effects of juvenile hormone on the programming of postembryonic development in eggs of the silkworm, *Hyalophora cecropia. Develop. Biol.*, **22**; 249–263.

Riddiford L. M. 1972. Juvenile hormone and insect embryonic development: Its potential role as an ovicide. [In:] *Insect Juvenile Hormones.* J. J. Menn and M. Beroza (eds.), pp. 95–111. Academic Press, New York.

Robbins W. E., Kaplanis J. N., Thompson M. J. Shortino T. J., Cohen C. F., Joyner S. C. 1968. Ecdysones and analogues: effects on development and reproduction of insects. *Science*, **161**; 1158–1159.

Rowlands D. G., Edwards J. P. 1986. New developments in pest control. [In:] *Spoilage and Mycotoxins of Cereals and other Stored Products.* B. Frannigan (ed.). *International Biodeterioration Suppl.*, **6**; 21–27.

Sehnal F., Meyer A. S. 1968. Larval-pupal transformation: Control by juvenile hormone, *Science*, **159**; 981–983.

Sehnal F. 1976. Action of juvenoids on different groups of insects. [In:] *The Juvenile Hormones*, L., J. Gilbert (ed.), pp. 301–322. Plenum Press, New York, London.

Sehnal F. 1985. Use of insect hormones, their analogues and anti-hormonal compounds in pest control. *Cesena Agricultura*, Cesena, pp. 223–236.

Sláma K., Williams C. M. 1965. Juvenile hormone activity for the bug *Pyrrhocoris apterus*. *Proc. Nat. Acad. Sci. USA.*, **54**; 411–414.

Sláma K., Williams C. M. 1966. "Paper factor" as an inhibitor of embryonic development of the European bug, *Pyrrhocoris apterus*. *Nature*, **210**; 329–330.

Sláma K. 1971. Insect juvenile hormone analogues. *Ann. Rev. Biochem.*, **40**; 1079–1102.

Sláma K., Romaňuk M., Šorm F. 1976. *Insect Hormones and Bioanalogues*. Springer-Verlag, Wien, New-York.

Šláma K., Romaňuk M. 1976. Juvenogens, biochemically activated juvenoid complexes. *Insect Biochem.*, **6**; 579–586.

Sláma K. 1978. The principles of antihormone action in insects. *Acta entomol. bohemoslov.*, **75**; 65–82.

Sláma K. 1980. Animal hormones and antihormones in plants. *Biochem. Physiol. Pflanzen* **175**; 177–193.

Staal G. B. 1975. Insect growth regulators with juvenile hormone activity. *Ann. Rev. Entomol.* **20**; 417–460.

Staal G. B. 1977. Insect control with insect growth regulators based on insect hormones. [In:] *Natural Products and the Protection of Plants*. G. B. Marini-Bettólo (ed.), pp. 253–283. Pontificia Academia Scientiarum, Roma.

Takahashi R. M., Miura T. 1975. Insect developmental inhibitors: Multiple applications of Dimilin and Altosid to *Gambusia affinis* (Baird and Girard). *Proc. and Papers 43rd Ann. Conf. Calif. Mosq. Cont. Assoc.*, 85–87.

Unnithan G. C., Nair K. K., Bowers W. S. 1977. Precocene-induced degeneration of the *corpus allatum* of adult females of the bug *Oncopeltus fasciatus*. *J. Insect Physiol.*, **23**; 1081–1094.

Vogel W., Masner P., Firschknecht M. L. 1976. The effect of the insect growth regulator epophenone on citrus mealbug and other homopterous insects. *XV. Int. Congr. Entomol. Washington D. C. USA, Aug.*, 19–27.

Wigglesworth V. B. 1963. The juvenile hormone effect of farnesol and some related compounds: Quantitative experiments. *J. Insect Physiol.*, **9**; 105–119.

Wigglesworth V. B. 1969. Chemical structure and juvenile hormone activity: Comparative tests on *Rhodnius prolixus*. *J. Insect Physiol.*, **15**; 73–94.

Williams C. M. 1956. The juvenile hormone of insects. *Nature*, **178**; 212–213.

Williams C. M. 1967. Third-generation pesticides. *Sci. Ann.*, **217**; 13–17.

Williams P., Amos T. G. 1974. Some effect of synthetic juvenile hormones and hormone analogues on *Tribolium castaneum* Aust. *J. Zool.*, **22**; 147–153.

Species index

Page numbers in bold type refer to figures

Subject index

Plates

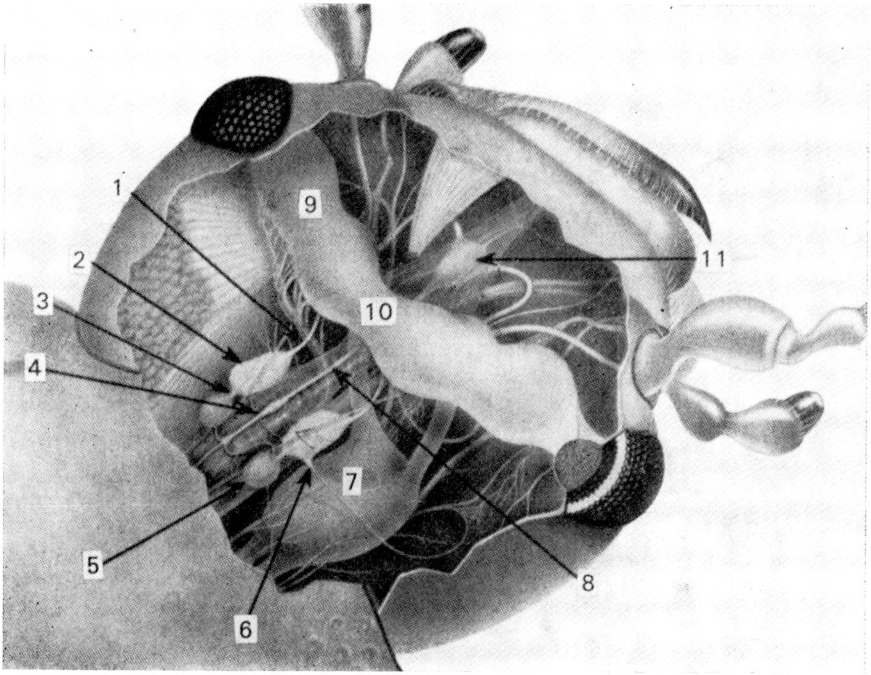

Plate 1. Neuroendocrine system in the head segment of the Colorado potato beetle (*Leptinotarsa decemlineata*). After de Wilde and Boer (1969).

1—*nervus corporis cardiaci*, 2—*corpus cardiacum*, 3—*nervus corporis allati*, 4— hypocerebral ganglion, 5—*corpus allatum*, 6—*nervus allatum-suboesophagealis*, 7—suboesophageal ganglion, 8—recurrent nerve, 9—optic lobe, 10—brain, 11—frontal ganglion.

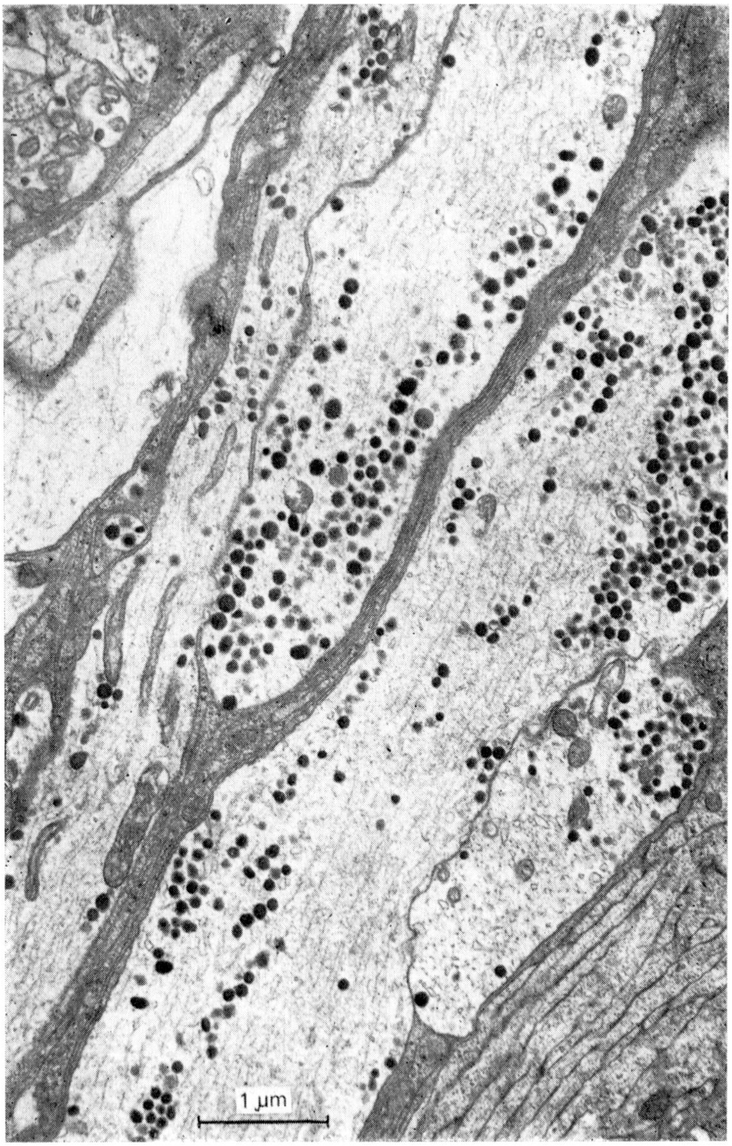

Plate 2. Electronogram of section of brain from last larval instar of the wax moth, showing the axons of neurosecretory cells containing numerous neurosecretory granules of different electron density and size. Photo. by S. Wartoń.

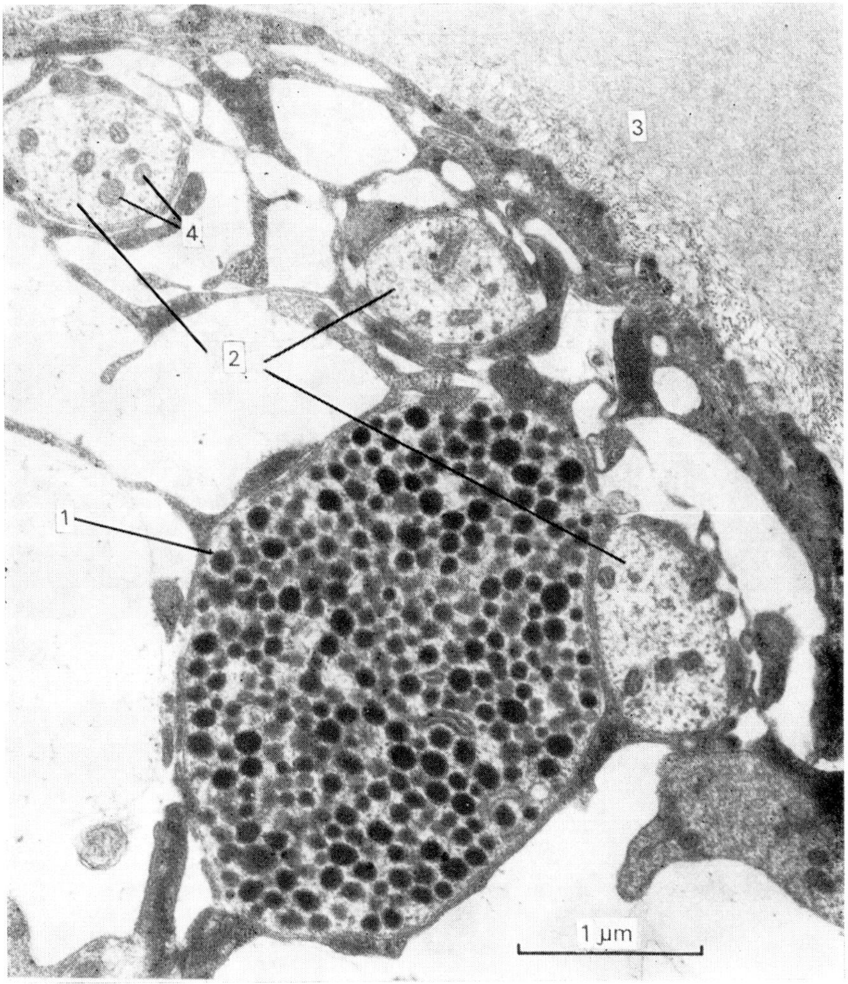

Plate 3. Electronogram of section through *nervi corporis cardiaci* of last-instar *Spodoptera littoralis* larva. Note axon of neurosecretory cell containing secretory granules of different sizes and electron densities (1), and three motor neuron axons (2). Photo. by S. J. Berry.

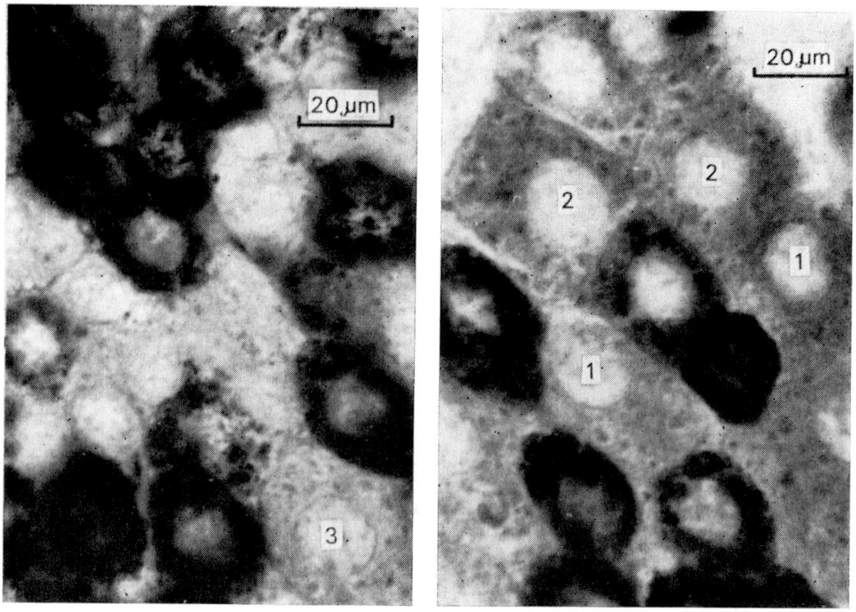

Plate 4. Various types of neurosecretory cells in brain of the cricket. Cells with dark-coloured cytoplasm are type A. Remaining cells are: 1—type *B*, 2—type *C*, 3—type *D*.

Plate 5. Electronogram of type *A* neurosecretory cell from brain of wax moth larva. Note neurosecretory granules of various sizes and electron densities. Photo. by S. Wartoń.

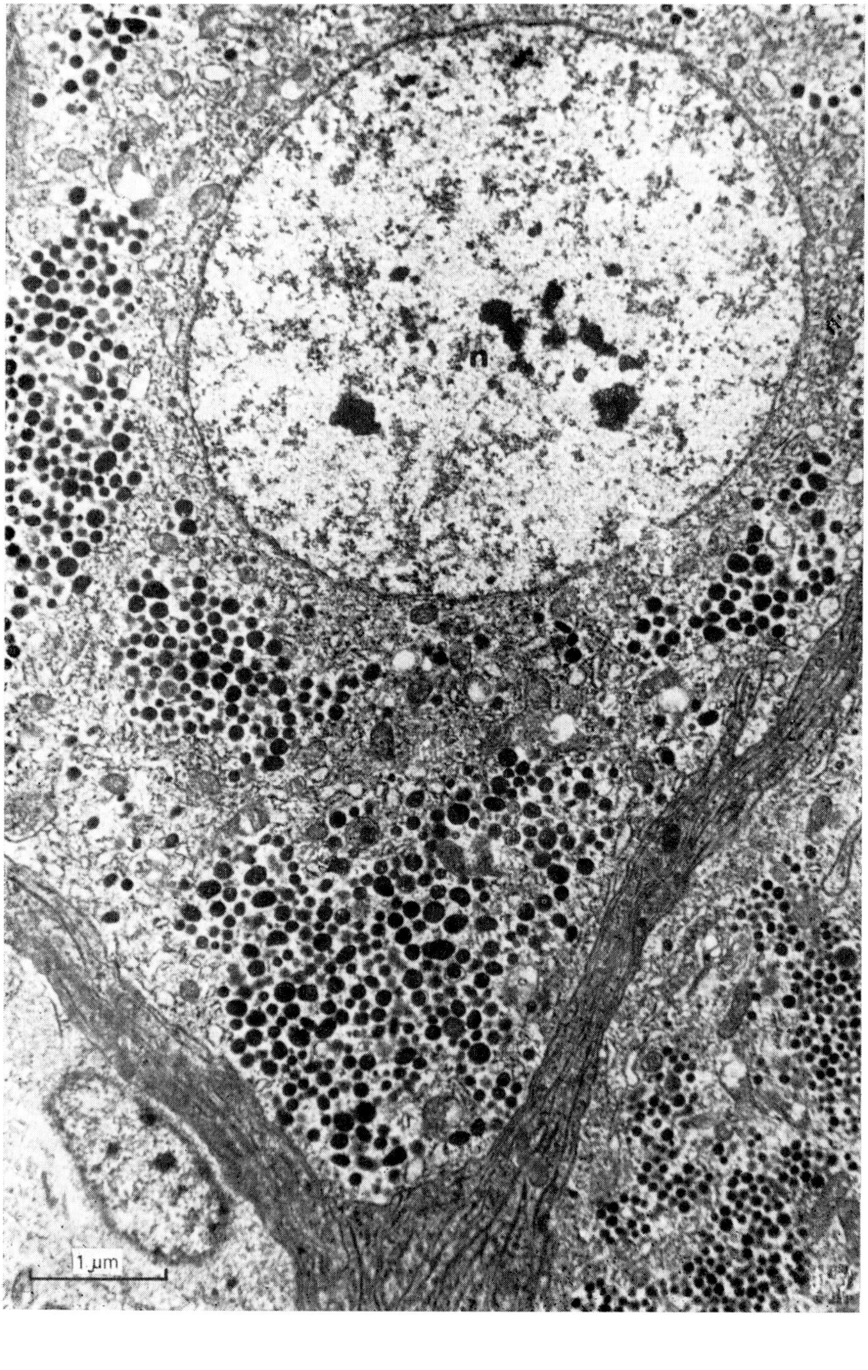
n
1 μm

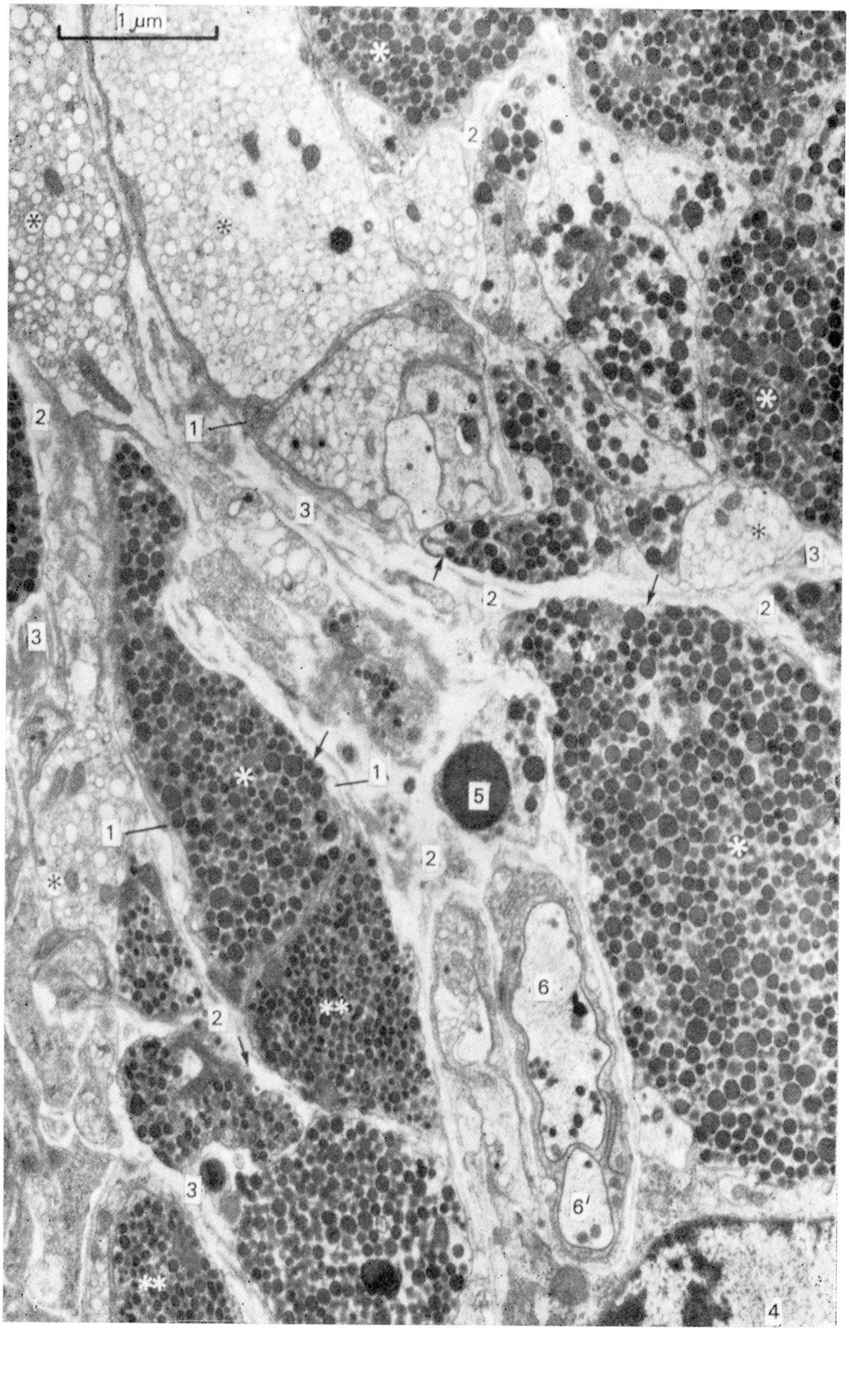
1 μm
2
1
3
3
2
2
3
1
1
5
2
2
6
6′
3
4

Plate 6. Electronogram of stick insect (*Carausius morosus*) *corpus cardiacum*. After Smith in Smith (1966).

Note large number of axon endings of neurosecretory cells filled with secretory granules, both electron-dense and clear (asterisks). In many places the membrane of these endings (1) is interrupted (arrows). It is at these sites that the axons join the intercellular spaces (2) which contain the material of which the basement membrane is composed (3) 4—glia cell nucleus, 5—large accumulation of neurosecretion, probably resulting from coalescence of numerous secretory granules, 6 and 6′—two small axons containing neurosecretory material.

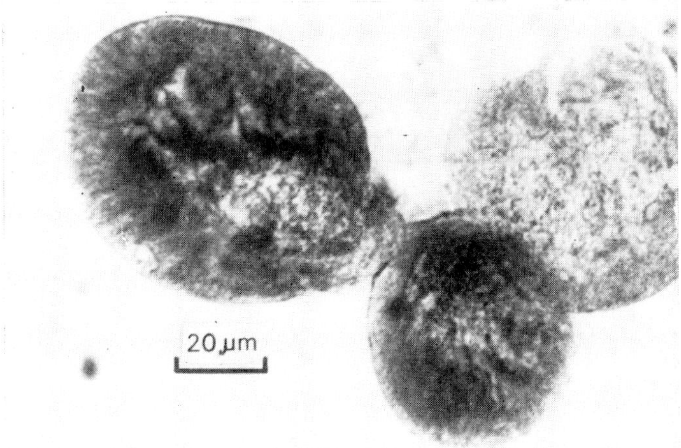

Plate 7. Histological section of prothoracic gland of *Spodoptera littoralis*. Note large nucleus of irregular shape occupying nearly the entire cell. Cytoplasm is visible only as a narrow band around the nucleus.

Plate 8. Method of placing ligature on final-instar larvae of wax moth (*Galleria mellonella*). A—ligature just posterior to head isolates retrocerebral cells: the brain, *corpora cardiaca*, *corpora allata*, frontal ganglion and suboesophageal ganglion, B—ligature posterior to thorax also isolates the prothoracic glands.

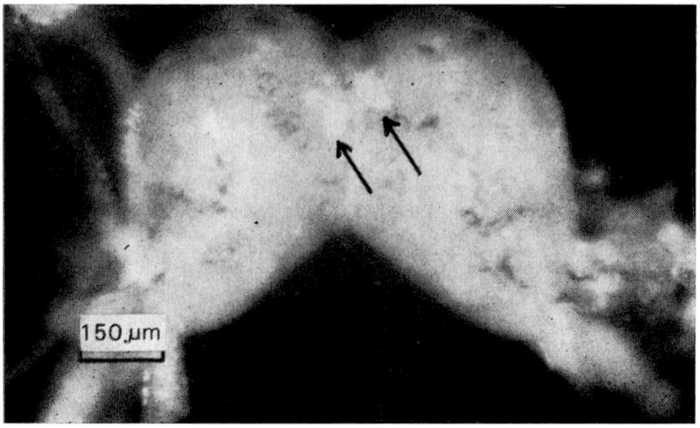

Plate 9. Distribution of neurosecretory cells in the brain of *Spodoptera littoralis* larva (arrows).

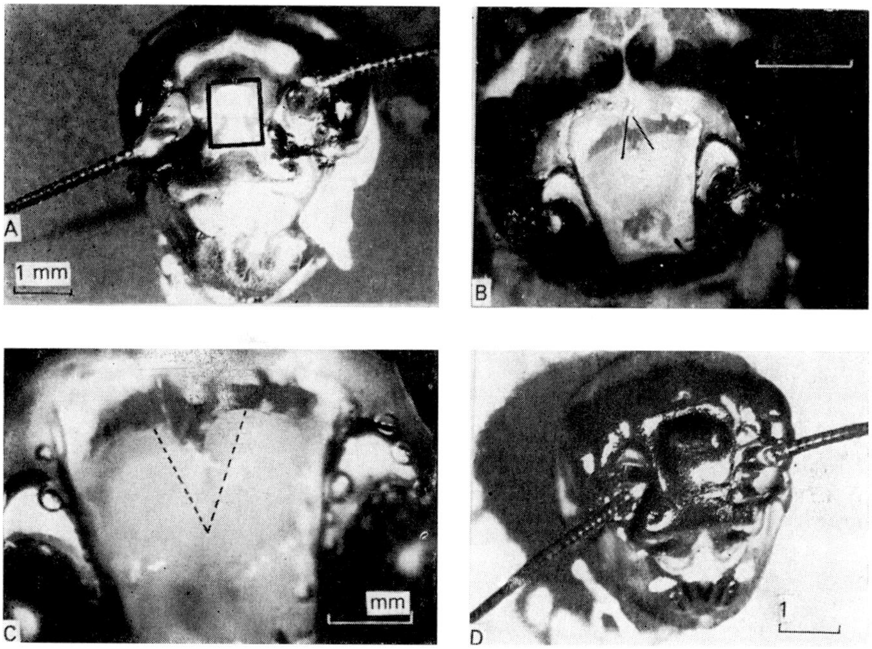

Plate 10. Stages in the removal of neurosecretory cells from the cricket (*Acheta domesticus*). A—position of opening in head capsule, B—brain visible after removal of cuticle, C—broken line indicates area in which neurosecretory cells of *pars intercerebralis* are located, D—head of insect after completion of operation and closure of opening by replacing cuticle using wax.

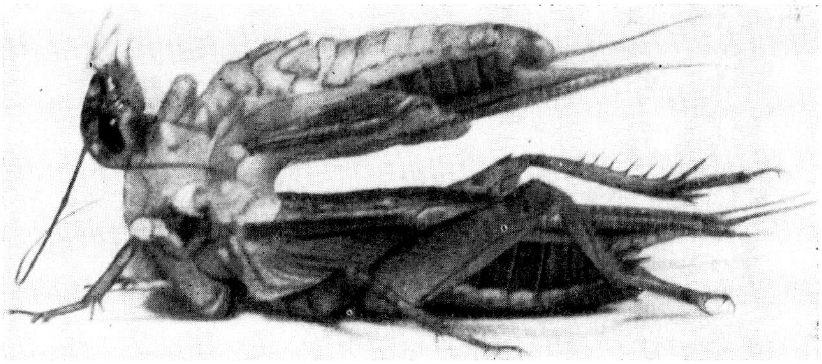

Plate 11. Parabiosis of crickets. The circulatory systems of the two insects are connected by a plastic tube. Further description in text.

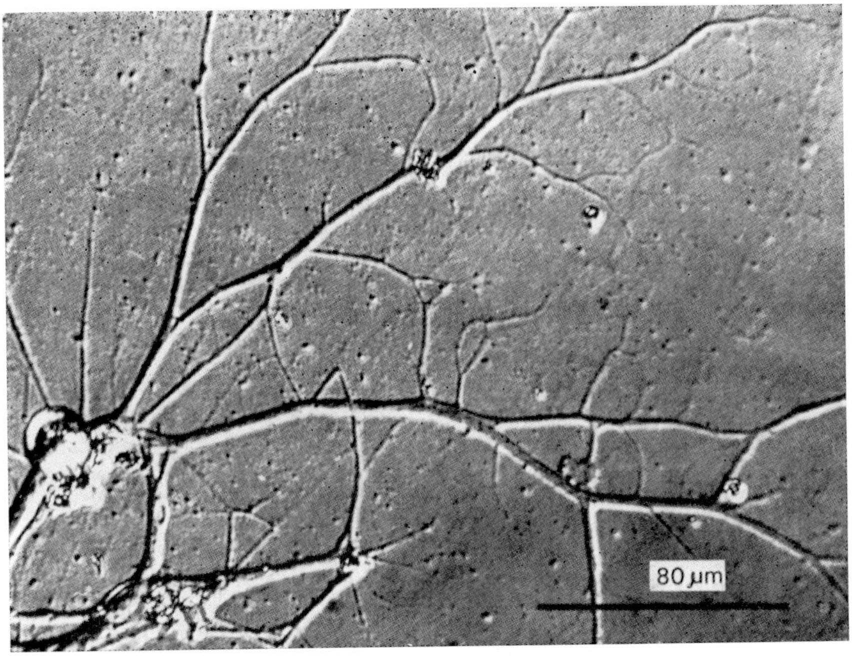
80 µm

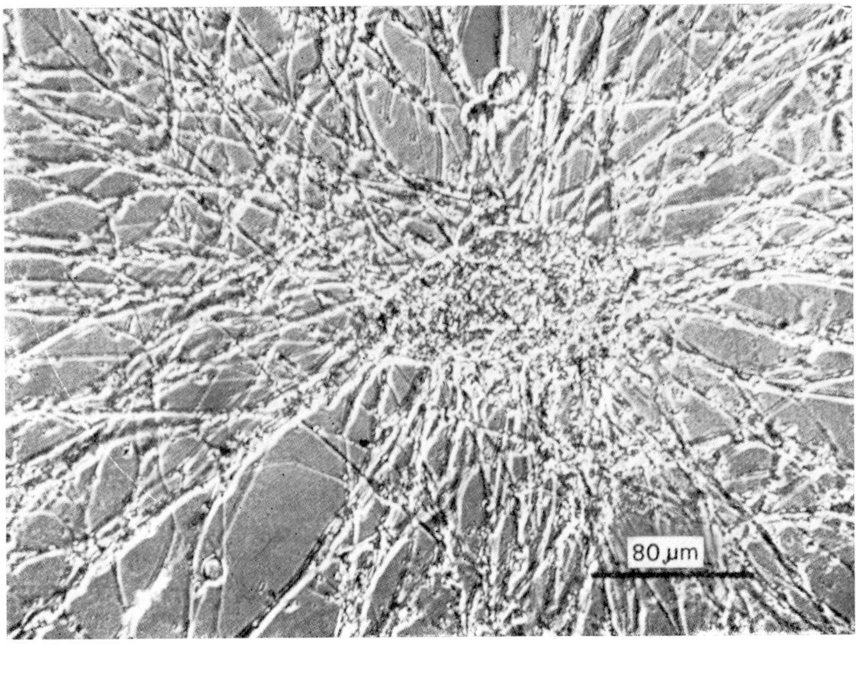
80 µm

Plate 12. Neurosecretory cell from brain of seventh-instar larva of cockroach, cultured for 33 days *in vitro*. Note the numerous processes of the cell. After Seshan (1976), with permission. Plate 13. *Corpus cardiacum* of adult cockroach, cultured for 47 days *in vitro*. Note the large number of processes formed by nerve cells. After Seshan (1976) with permission.

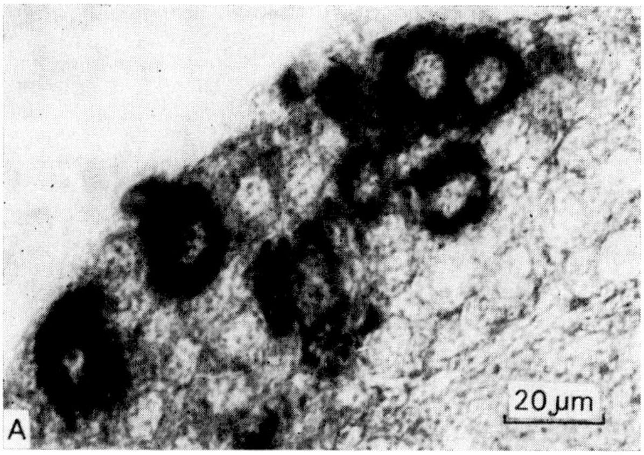

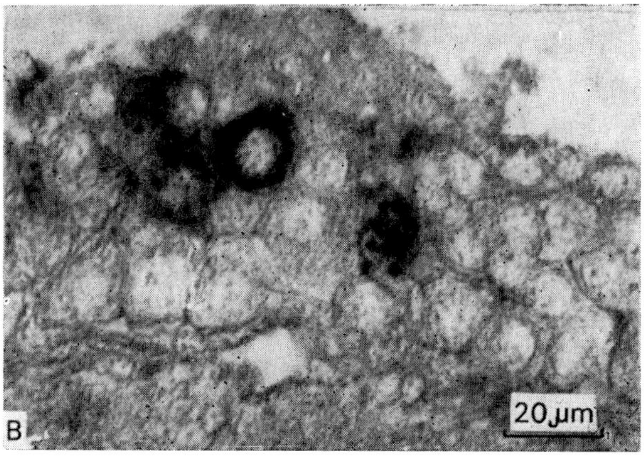

Plate 14. Neurosecretory cells of *pars intercerebralis* of cricket brain (A). Note large number of cells completely filled with neurosecretory material. Neurosecretory cells of suboesophageal ganglion of the same insect (B) in various stages of accumulation of neurosecretory material. Cells containing single neurosecretory granules are visible. Preparations fixed in Bouin's solution and stained with aldehyde fuchsin.

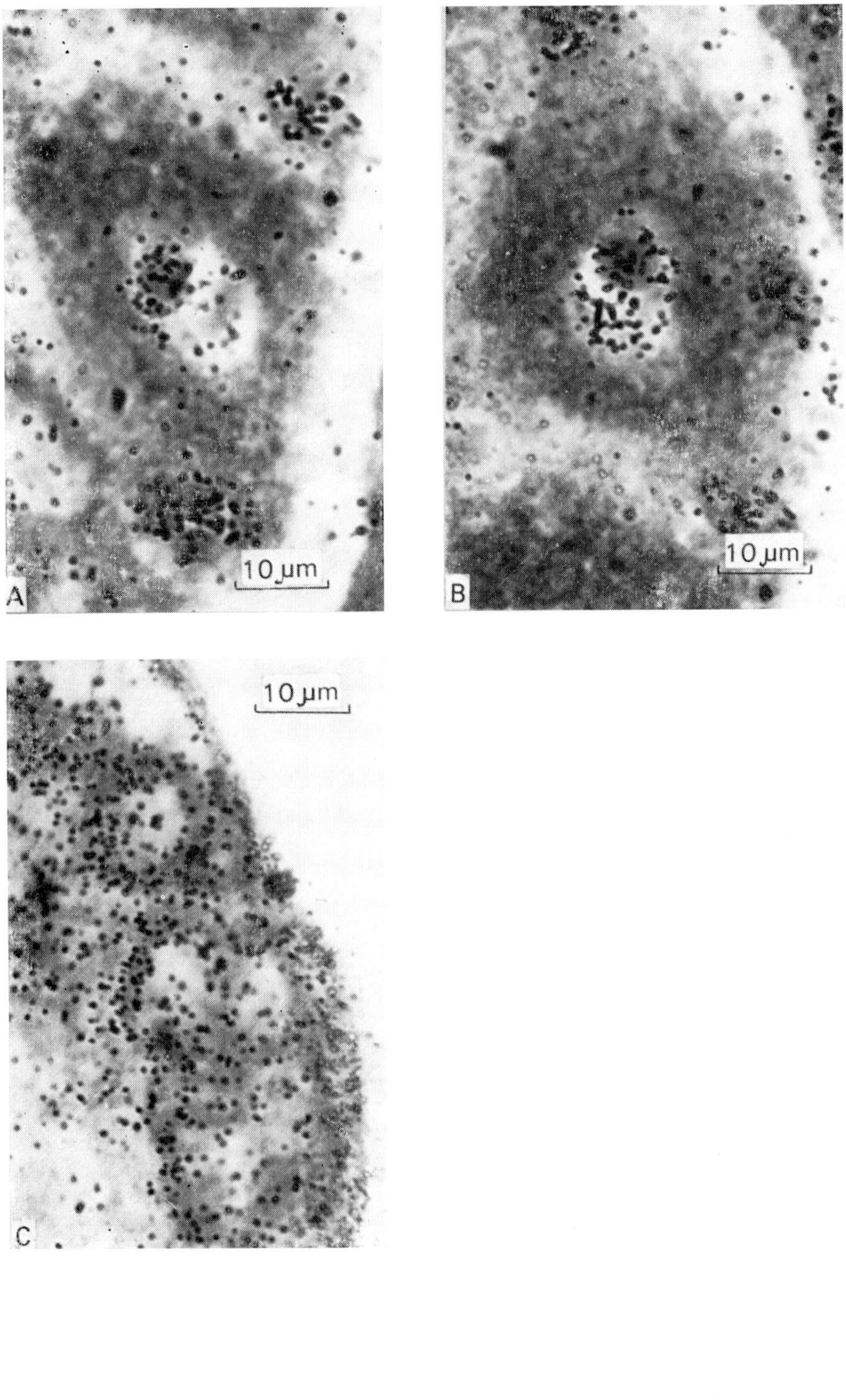
10 μm
A
10 μm
B
10 μm
C

Plate 15. Autoradiograms of neurosecretory cells of cricket brains. Note varying distribution of RNA labelled with ^{3}H-uridine. Incorporation of labelled uridine into the nucleolus (A), nucleolus and nucleus (B) and the cytoplasm (C) is shown. Exposure time was 6 weeks. Autoradiogram stained with methyl green and pyronin.

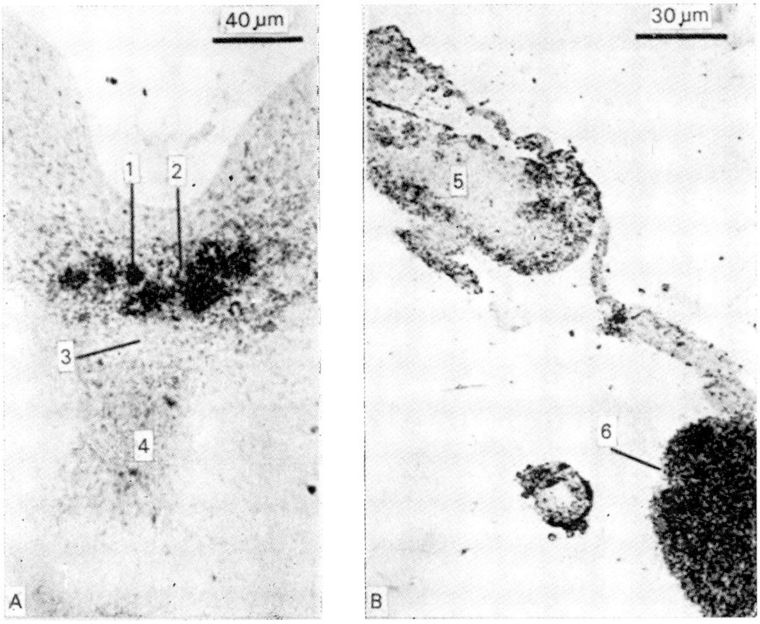

Plate 16. Autoradiogram of neurosecretory cells of *pars intercerebralis* of the cricket brain (A) and of *corpora cardiaca* and *corpora allata* (B). The autoradiograms were obtained by injecting ^{36}S-cysteine into the insects for two hours. Note the presence of labelled protein in type *A* (1) and type *B* (2) neurosecretory cells and in the axons of type *C* neurosecretory cells (3). After Cymborowski (1973).

4—*nervus corporis cardiaci I*, 5—*corpus cardiacum*, 6—*corpus allatum*.

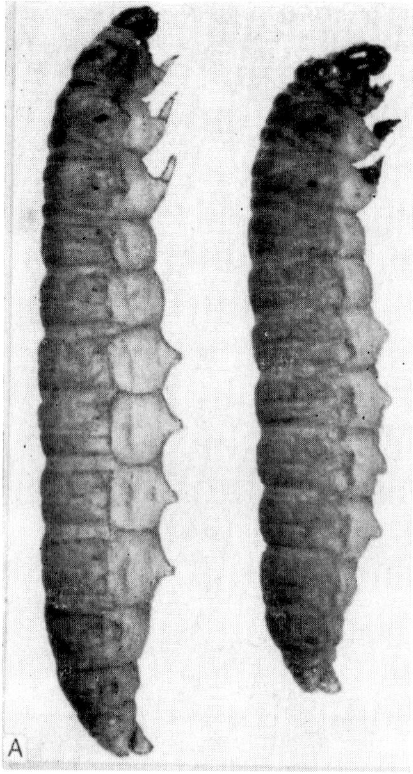
A

B

C

D

Plate 17. Morphological changes in final-instar (VII) larvae of *Galleria mellonella*.
A (left) superlarva (stage VIII), grade 5 morphological changes (see Table 1), A (right) larva with grade 4 morphological changes, VII stage. B—forms intermediate between larva and pupa—grade 3. C—pupa with larval cuticle (arrows). D—grade 2 morphological changes. D (left)—pupa with short wing discs, grade 1 changes. D (right)—normal pupa.

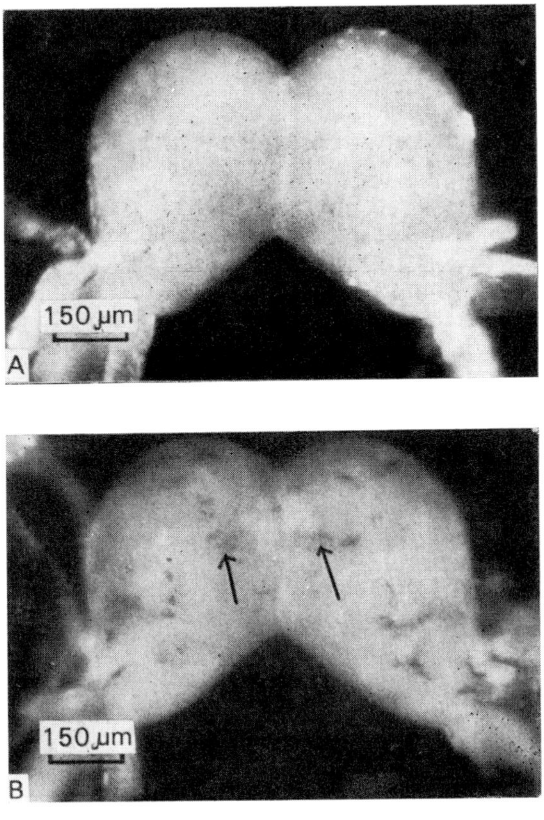

Plate 18. Changes in pigment distribution in brain of *Spodoptera littoralis* during last larval stage.
A—brain of newly formed final-instar larva. Note uniform distribution of pigment in brain, B—brain of larva before pupation; pigment has concentrated, some pigment cells are in close contact with neurosecretory cells (arrows).

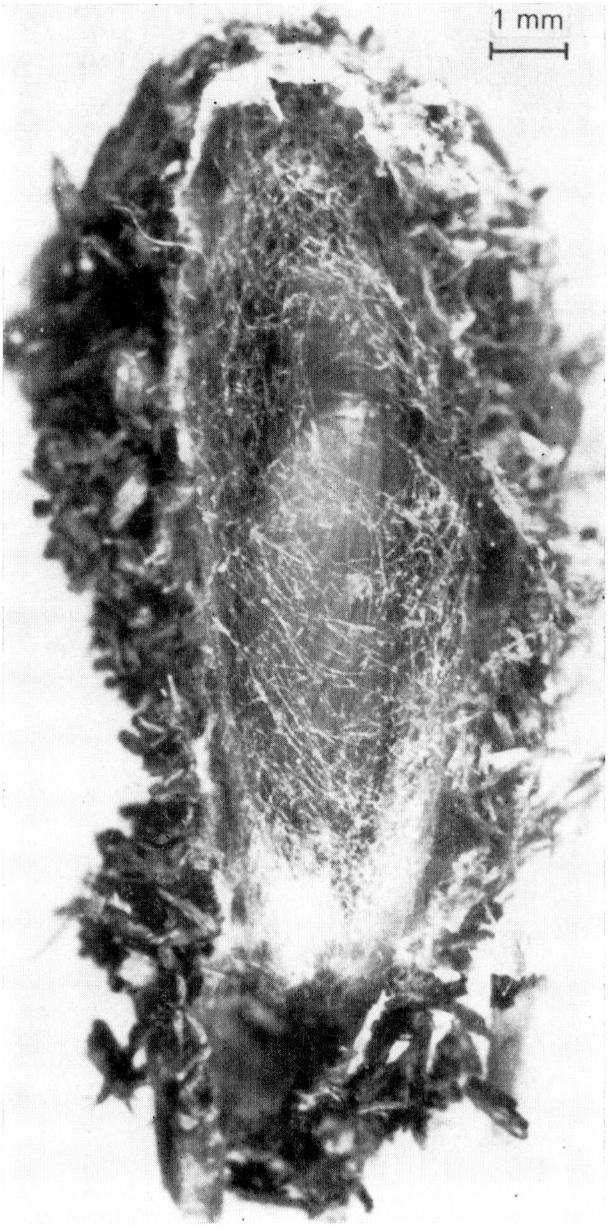

Plate 19. Structure of normal cocoon of final-instar *Ephestia kuehniella* larvae before metamorphosis. Note two layers: outer layer may contain material gathered by the larva in addition to silk; inner layer of silk only. After Giebułtowicz (1980) with permission.

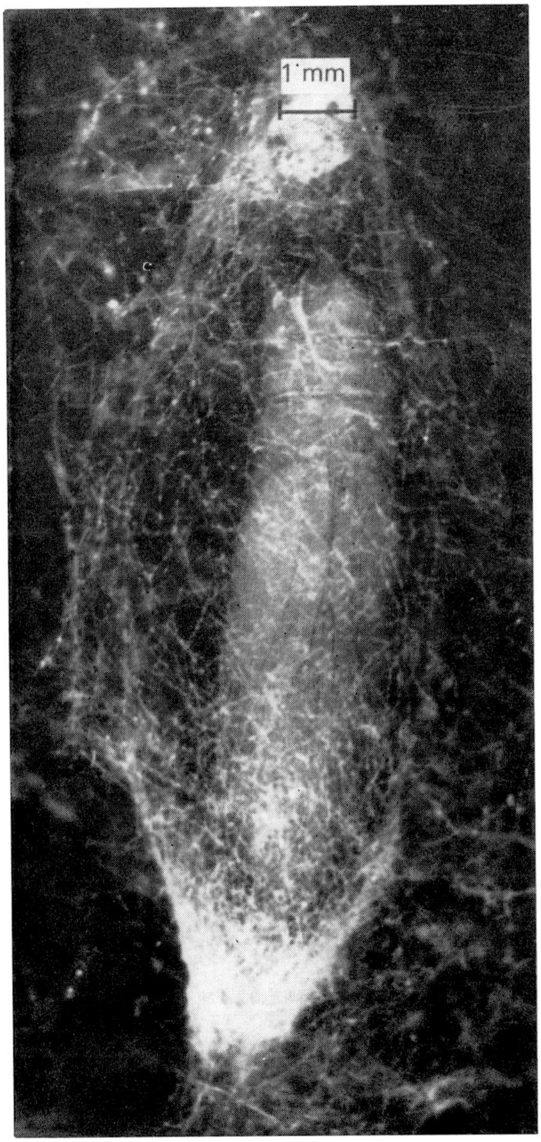

Plate 20. Pupation of *Ephestia kuehniella* larva that has produced only the inner cocoon layer. After Giebułtowicz (1980) with permission.

Plate 21. Comparison of morphological structure of *Galleria mellonella* larva in last instar (above) with superlarva obtained by treatment with juvenoid (below).

Plate 22. Effect of juvenoids on metamorphosis of *Dysdercus*. From right to left: final-instar larva, normal adult, form intermediate between larva and adult (adultoid), superlarva. Photo. by K. Sláma.